5

COLLOQUIA

Enrico Bombieri
School of Mathematics
Institute for Advanced Study
Einstein Drive
Princeton, NJ 08540 USA

Pierre Cartier
Institut des Hautes Études Scientifiques
35, Route de Chartres
F–91440 Bures-sur-Yvette, France

Fabrizio Catanese
Mathematisches Institut
Lehrstuhl Mathematik VIII
Universitätsstr. 30
95447 Bayreuth, Germany

John Brian Conrey
American Institute of Mathematics
360 Portage Ave
Palo Alto, CA 94306, USA

Alessio Figalli
Department of Mathematics
The University of Texas at Austin
2515 Speedway Stop C1200
Austin TX 78712, USA

Étienne Fouvry
Université Paris-Sud
Laboratoire de Mathématique
Campus d'Orsay
91405 Orsay Cedex, France

Emmanuel Kowalski
ETH Zürich – Departement Mathematik
Rämistrasse 101
CH-8092 Zürich, Switzerland

Luc Illusie
Mathématique, Bât. 425
Université Paris-Sud
91405 Orsay Cedex, France

Philippe Michel
EPFL/SB/IMB/TAN
Station 8
CH-1015 Lausanne, Switzerland

Michel Waldschmidt
Institute de Mathématiques de Jussieu
Université Pierre et Marie Curie Paris 6
4 Place Jussieu
75252 Paris, France

Colloquium De Giorgi 2013 and 2014

Colloquium De Giorgi 2013 and 2014

edited by
Umberto Zannier

EDIZIONI
DELLA
NORMALE

ISBN 978-88-7642-514-1 ISBN 978-88-7642-515-8 (eBook)
DOI 10.1007/978-88-7642-515-8

Contents

Preface

Since 2001 the Scuola Normale Superiore di Pisa has organized the "Colloquio De Giorgi", a series of colloquium talks named after Ennio De Giorgi, the eminent analyst who was a Professor at the Scuola from 1959 until his death in 1996.

The Colloquio takes place once a month. It is addressed to a general mathematical audience, and especially meant to attract graduate students and advanced undergraduate students. The lectures are intended to be not too technical, in fields of wide interest. They should provide an overview of the general topic, possibly in a historical perspective, together with a description of more recent progress.

The idea of collecting the materials from these lectures and publishing them in annual volumes came out recently, as a recognition of their intrinsic mathematical interest, and also with the aim of preserving memory of these events.

For this purpose, the invited speakers are now asked to contribute with a written exposition of their talk, in the form of a short survey or extended abstract. This series has been continued in a collection that we hope shall be increased in the future.

This volume contains a complete list of the talks held in the "Colloquio De Giorgi" in recent years but also in the past and also in the past years, and a table of contents of all the volumes too.

Colloquia held in 2001

Paul Gauduchon
Weakly self-dual Kähler surfaces

Tristan Rivière
Topological singularities for maps between manifolds

Frédéric Hélein
Integrable systems in differential geometry and Hamiltonian stationary Lagrangian surfaces

Jean-Pierre Demailly
Numerical characterization of the Kähler cone of a compact Kähler manifold

Elias Stein
Discrete analogues in harmonic analysis

John N. Mather
Differentiability of the stable norm in dimension less than or equal to three and of the minimal average action in dimension less than or equal to two

Guy David
About global Mumford-Shah minimizers

Jacob Palis
A global view of dynamics

Alexander Nagel
Fundamental solutions for the Kohn-Laplacian

Alan Huckleberry
Incidence geometry and function theory

Colloquia held in 2002

Michael Cowling
Generalizzazioni di mappe conformi

Felix Otto
The geometry of dissipative evolution equations

Curtis McMullen
Dynamics on complex surfaces

Nicolai Krylov
Some old and new relations between partial differential equations, stochastic partial differential equations, and fine properties of the Wiener process

Tobias H. Colding
Disks that are double spiral staircases

Cédric Villani
When statistical mechanics meets regularity theory: qualitative properties of the Boltzmann equation with long-range interactions

Colloquia held in 2003

John Toland
Bernoulli free boundary problems-progress and open questions

Jean-Michel Morel
The axiomatic method in perception theory and image analysis

Jacques Faraut
Random matrices and infinite dimensional harmonic analysis

Albert Fathi
C^1 *subsolutions of Hamilton-Iacobi Equation*

Hakan Eliasson
Quasi-periodic Schrödinger operators-spectral theory and dynamics

Yakov Pesin
Is chaotic behavior typical among dynamical systems?

Don B. Zagier
Modular forms and their periods

David Elworthy
Functions of finite energy in finite and infinite dimensions

Colloquia held in 2004

Jean-Christophe Yoccoz
Hyperbolicity for products of 2×2 *matrices*

Giovanni Jona-Lasinio
Probabilità e meccanica statistica

John H. Hubbard
Thurston's theorem on combinatorics of rational functions and its generalization to exponentials

Marcelo Viana
Equilibrium states

Boris Rozovsky
Stochastic Navier-Stokes equations for turbulent flows

Marc Rosso
Braids and shuffles

Michael Christ
The d-bar Neumann problem, magnetic Schrödinger operators, and the Aharonov-Böhm phenomenon

Colloquia held in 2005

Louis Nirenberg
One thing leads to another

Viviane Baladi
Dynamical zeta functions and anisotropic Sobolev and Hölder spaces

Giorgio Velo
Scattering non lineare

Gerd Faltings
Diophantine equations

Martin Nowak
Evolution of cooperation

Peter Swinnerton-Dyer
Counting rational points: Manin's conjecture

François Golse
The Navier-Stokes limit of the Boltzmann equation

Joseph J. Kohn
Existence and hypoellipticity with loss of derivatives

Dorian Goldfeld
On Gauss' class number problem

Colloquia held in 2006

Yuri Bilu
Diophantine equations with separated variables

Corrado De Concini
Algebre con tracce e rappresentazioni di gruppi quantici

Zeev Rudnick
Eigenvalue statistics and lattice points

Lucien Szpiro
Algebraic Dynamics

Simon Gindikin
Harmonic analysis on complex semisimple groups and symmetric spaces from point of view of complex analysis

David Masser
From 2 to polarizations on abelian varieties

Colloquia held in 2007

Klas Diederich
Real and complex analytic structures

Stanislav Smirnov
Towards conformal invariance of 2D lattice models

Roger Heath-Brown
Zeros of forms in many variables

Vladimir Sverak
PDE aspects of the Navier-Stokes equations

Christopher Hacon
The canonical ring is finitely generated

John Coates
Elliptic curves and Iwasava theory

Colloquia held in 2008

Claudio Procesi
Funzioni di partizione e box-spline

Pascal Auscher
Recent development on boundary value problems via Kato square root estimates

Hendrik W. Lenstra
Standard models for finite fields

Jean-Michel Bony
Generalized Fourier integral operators and evolution equations

Shreeram S. Abhyankar
The Jacobian conjecture

Fedor Bogomolov
Algebraic varieties over small fields

Louis Nirenberg
On the Dirichlet problem for some fully nonlinear second order elliptic equations

Colloquia held in 2009

Michael G. Cowling
Isomorphisms of the Figa-Talamanca-Herz algebras $Ap(G)$ for connected Lie groups G

Joseph A. Wolf
Classical analysis and nilpotent Lie groups

Gisbert Wustholz
Leibniz' conjecture, periods & motives

David Mumford
The geometry and curvature of shape spaces

Colloquia held in 2010

Charles Fefferman
Extension of functions and interpolation of data

Colloquia held in 2011

Ivar Ekeland
An inverse function theorem in C^{∞}

Pierre Cartier
Numbers and symmetries: the 200th anniversary of Galois' birth

Yves André
Galois theory beyond algebraic numbers and algebraic functions

Colloquia held in 2012

Nicholas M. Katz
Simple things we don't know

Endre Szemeredi
Is laziness paying off? ("Absorbing" method)

Dan Abramovich
Moduli of algebraic and tropical curves

Elias M. Stein
Three projection operators in complex analysis

Shou-Wu Zhang
Congruent numbers and Heegner points

Enrico Bombieri
The Mathematical Truth

Colloquia held in 2013

Pierre Cartier
New developments in Galois theory

Étienne Fouvry, Emmanuel Kowalski and Philippe Michel
Trace functions over finite fields and their applications

Fabrizio Catanese
Topological methods in algebraic geometry

Luc Illusie
Grothendieck at Pisa: crystals and Barsotti-Tate groups

Brian Conrey
Riemann's hypothesis

Alessio Figalli
Stability results for the Brunn-Minkowski inequality

Colloquia held in 2014

Michel Waldschmidt
Schanuel's Conjecture: algebraic independence of transcendental numbers

Contents of previous volumes:

Colloquium De Giorgi 2006

Colloquium De Giorgi 2007/2008

Colloquium De Giorgi 2009

Colloquium De Giorgi 2010-2012

The Mathematical Truth[1]

Enrico Bombieri

The aim of this lecture is to give an overview of several different notions of the concepts of truth and proof in mathematics. This includes the two main directions of 'Platonic realism' and 'Formalism', with some variants, and other views such as 'Intuitionism', empiricism and quasi-empiricism, Field's fictionalism, and social constructivism and realism. The lecture concludes with remarks on the notion of proof, including very recent progress obtained by computer scientists for understanding the overall notion of complexity of proof checking, and finally with some personal reminiscences and remarks on the subject.

The aim of this lecture was to give an overview of several different notions of the concepts of truth and of the related concept proof in mathematics. The notion of truth has many different facets and has been extensively treated by philosophers, theologians, historians, lawyers, and mathematicans as well. The layman views truth as an absolute concept; the theologian distinguishes between rational truth and theological truth; philosophers, talk about rational truth and truth by consensus; for a lawyer, the ultimate level of justice becomes established truth. The mathematician dstinguishes between verifiable truth and unverifiable truth, a delicate concept that is determined by the logical system in which one may be working.

The Platonic view ("Platonic realism") of an universe of ideas and absolute truths has been accepted only with many caveats by mathematicians. For many, there is no longer a single universe of mathematical ideas, instead the mathematical universe is determined by a small set of logical rules ("the axioms") and it consists of the non-contradictory sentences of the language determined by the axioms ("the formalistic model").

[1] The full text of this lecture has been published in: *The shifting aspects of truth in mathematics*, Enrico Bombieri, Euresis Journal, Volume 5, Summer 2013, 249–272.

The naive view that identifies truth with verification or proof has been changing quite a bit in the mathematical world. Although truth is the glue that holds together mathematics, most mathematicians believe that mathematics must be viewed as coherent collections of truths, or probable truths ("the conjectures") which have much more meaning when considered as a single object rather than of many disjoint parts. This may be understood in the same way that a beautiful painting is much more than a collection of colored pigments on a canvas. Philosophers such as Wittgenstein have also embraced this vision, as a fundamental setting.

In view of the importance of truth in mathematics and the difficulty that one may encounter in the process of verification, many questions arise.

1) Is classical mathematics free from contradiction?
2) Is truth identifiable with verification?
3) Can truth, or proof be achieved by consensus?
4) Is there a mathematical notion of probable truth?
5) Is automatic verification (for example, by computer) acceptable in mathematics?

The surprising answer to question 1) was obtained by Tarski: Any sufficiently large non-contradictory model of mathematics cannot prove its consistency within itself. To this one can add Gödel proof that in any large non-contradictory system there exist true statements that are unprovable within the system itself.

Question 3), 4), and 5) are interrelated. Question 3) has ramifications into aspects of the social sciences. Question 4) is related to the concept of peer review which is so basic today for the progress of science in general, but also can be put in a precise formal setting that can be analyzed with mathematical precision by a computer program. Finally, question 5) is becoming of great actuality today because of new logical programs available to computers that can quickly do the formal verification of extremely complex mathematical papers. There are now deep mathematical results that have been "proved" only by using many hours of computing in order to deal with extremely long or complicated numerical calculations unfeasible by hand.

The lecture concludes with the view that truth in mathematics is not irrelevant nor tautological, it is the glue that hold the fabric of mathemaatics together. Only with adherence to truth mathematics can maintain its intellectual attraction and its importance in human history.

New developments in Galois theory

Pierre Cartier

1. A return to Galois methods

We consider a polynomial

$$P(X) = X^n + a_1 X^{n-1} + \cdots + a_n. \tag{1.1}$$

The coefficients $a_1, \ldots, a_n$ should belong to some field K, called by Galois a domain of rationality. We are seeking roots for P in a larger field Ω. We may suppose that Ω is large enough to allow a complete decomposition

$$P(X) = (X - r_1) \cdots (X - r_n); \tag{1.2}$$

for instance, if Ω is algebraically closed, every polynomial admits such a decomposition. To every such a decomposition we associate a point $r = (r_1, \ldots, r_n)$ in Ω^n. The collection of such points is a finite set $\mathcal{S}$, closed for the K–Zariski topology of Ω^n (a paraphrase of some of Galois definitions). Moreover the symmetric group Σ_n of order $n!$ acts on $\mathcal{S}$.

The Zariski topology on $\mathcal{S}$ gives rise to a partition of $\mathcal{S}$ into the minimal closed subsets of $\mathcal{S}$. Call Π the set of "blocks", on which Σ_n acts transitively. In intrinsic way, such an action defines a groupoid T, which should be called the *Galois groupoid* of the equation $P = 0$. By selecting a block b in Π, we obtain a subgroup of Σ_n, defined up to conjugation, and this is the *Galois group*. In interesting situations already known to Galois, the Galois group is the group of symmetries of a finite geometry; for instance, the Galois group of the equation $X^7 - 7X + 3 = 0$ with $K = \mathbb{Q}$ is the simple group of order 168, the group of symmetries of a finite projective plane whose 7 points correspond to the 7 roots.

2. Differential Galois theory

There is another instance where groupoids are at the heart of Galois theory, namely differential equations. Let us consider for instance a plane with coordinates x, y and the differential form $\omega = dy - ydx$. By taking y as an independent variable, the equation $\omega = 0$ takes the form $dx = dy/y$ with solution $x = \log y$. But in the complex domain, the logarithmic function is not single–valued, with branches $2\pi ik + \log y$ ($k \in \mathbb{Z}$). There is a natural group of translations on these branches, namely the group $2\pi i\mathbb{Z}$ of integral multiples of $2\pi i$. The smallest algebraic subgroup containing $2\pi i\mathbb{Z}$ is the full group $\mathbb{G}_a$ of translations of the line. The *ambiguity* on the logarithm expressed by the monodromy ($\log y \mapsto 2\pi i + \log y$) is encoded in the *differential Galois group* $\mathbb{G}_a$.

Take now x as independent variable: we get the linear differential equation

$$\frac{dy}{dx} = y \tag{2.1}$$

with solution $y = \exp(x)$. According to the classical Picard–Vessiot theory, to such a linear system of order n

$$\frac{d\vec{y}}{dx} = A(x)\vec{y} \tag{2.2}$$

($\vec{y} = (y_1, \ldots, y_n)$ is a vector) is associated a certain algebraic subgroup of $\mathrm{GL}(n)$, the Galois–Picard–Vessiot group. In our example, the group is $\mathrm{GL}(1)$ (that is the multiplicative group $\mathbb{G}_m$ of nonzero numbers).

As Lie groups of dimension 1, the groups $\mathbb{G}_a$ and $\mathbb{G}_m$ are locally isomorphic (either in the real or the complex domain). *As algebraic groups, they are definitely non–isomorphic*. What is the true Galois group?

A recent answer originated in the works of Ramis and Malgrange, inspired by the return map of Poincaré in dynamical systems. Viewed geometrically, the differential equation $\omega = 0$ (that is $dy = ydx$) corresponds to a *foliation* whose leaves are the integral curves $y = c\exp x$. We have the notion of a local transversal to such a foliation, a little arc T crossing non tangentially the integral curves C

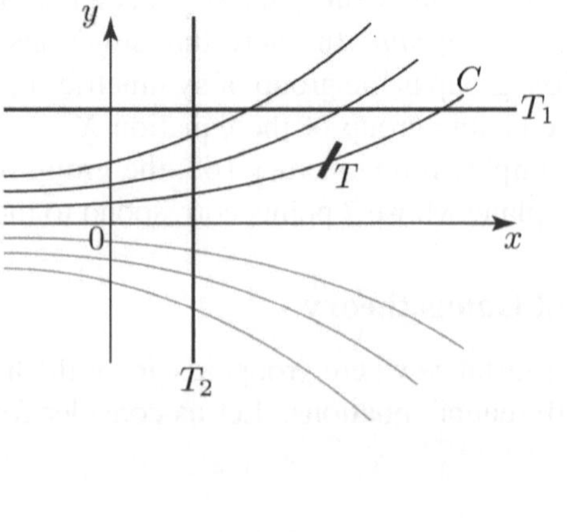

According to É. Cartan (the father), we have a groupoid (or rather a pseudo–group) of transformations of these local transversals. Inside this pseudo–group, there is an *algebraic pseudo–group* corresponding to the *holonomy* of the foliation defined by $\omega = 0$. The pseudo–group unifies the two previous Galois groups $\mathbb{G}_a$ and $\mathbb{G}_m$, corresponding to two different *global transversals* T_1 and T_2.

3. Periods

Interesting numbers can be defined by integrals, namely

$$\log 2 = \int_1^2 \frac{dx}{x}, \qquad \pi = 4\int_0^1 \frac{dx}{x^2+1}, \qquad 2\pi i = \oint \frac{dz}{z}$$

where the last integral corresponds to a closed loop around 0 in $\mathbb{C}$. Other interesting numbers are the zeta–values

$$\zeta(k) = \sum_{n\geq 1} \frac{1}{n^k} \tag{3.1}$$

for $k \geq 2$. Again they are expressed as integrals, namely

$$\zeta(k) = \int_{[0,1]^k} \frac{dx_1 \cdots dx_k}{1 - x_1 \cdots x_k}. \tag{3.2}$$

Galois theory deals with algebraic numbers, but the previous numbers are known or expected to be transcendental. *Can we have a kind of Galois theory, not for all transcendental numbers, but for some of the previous kind?*

The first attempt was made by Kontsevich and Zagier. They considered under the name of *periods* a certain class of definite integrals (generalizing (3.2) for instance). Now we consider a number of operations dealing with formulas without resorting to limit procedures. For instance, we can transform

$$\zeta(3) = \int_0^1 \int_0^1 \int_0^1 \frac{dxdydz}{1-xyz} \quad \text{into} \quad \int_\Delta \frac{dudvdw}{(1-u)vw},$$

where the domain Δ is given by the conditions $0 \leq u \leq v \leq w \leq 1$, by using the change of variables

$$u = xyz, \qquad v = yz, \qquad w = z.$$

In this way one creates a commutative algebra $\mathcal{P}$ over the field $\mathbb{Q}$ of rational numbers, together with a homomorphism $ev : \mathcal{P} \to \mathbb{C}$ (think of elements of $\mathcal{P}$ as equivalence classes of programs and *ev* as the "exec" command). It should be expected that *ev* is an injective map. The meaning is that once you have proven the famous Euler relation $\zeta(2) = \pi^2/6$ by whatever means, there should exist an elementary proof for it.

The original proposal has been revised by Deligne, Goncharov, and lately by Francis Brown. As a result we have an abstract algebra $\mathcal{P}^{\mathfrak{m}}$ of *multizeta motivic periods*, and an evaluation map $ev^{\mathfrak{m}} : \mathcal{P}^{\mathfrak{m}} \to \mathbb{C}$. Not only the zeta–values $\zeta(k)$ can be lifted to elements $\zeta^{\mathfrak{m}}(k)$ in $\mathcal{P}^{\mathfrak{m}}$, but also the Euler double zeta–values like

$$\zeta(a, b) = \sum_{0<m<n} \frac{1}{m^a n^b}, \tag{3.3}$$

and more generally multizeta values. We control the automorphism group of $\mathcal{P}^{\mathfrak{m}}$, properly called *the motivic Galois group*. We expect the evaluation map from $\mathcal{P}^{\mathfrak{m}}$ to $\mathbb{C}$ to be injective. This would imply for instance that the numbers

$$\pi, \ \zeta(3), \ \zeta(5) \ldots$$

are transcendental and algebraically independent (a wild dream to be fulfilled around 2040!). But it could provide a large group of symmetries of these numbers: a truly Galois group for transcendental numbers!

Trace functions over finite fields and their applications

Étienne Fouvry, Emmanuel Kowalski and Philippe Michel

Abstract. We survey our recent works concerning applications to analytic number theory of trace functions of ℓ-adic sheaves over finite fields.

1. Motivation

We begin by describing one of the motivating problems for our paper [7]. This concerns an equidistribution statement in the upper half-plane $\mathbf{H}$ of complex numbers with positive imaginary parts, or more precisely in the domain

$$F = \{z \in \mathbf{H} \mid |\mathrm{Re}(z)| \leq 1/2,\ |z| \geq 1\} \subset \mathbf{H}.$$

This closed subset of $\mathbf{H}$ is well-known to be a *fundamental domain* for the action of the modular group $\mathrm{SL}_2(\mathbf{Z})$ by homographies on $\mathbf{H}$, *i.e.*, the restriction to $\mathrm{SL}_2(\mathbf{Z})$ of the $\mathrm{SL}_2(\mathbf{R})$-action given by

$$\begin{pmatrix} a & b \\ c & d \end{pmatrix} \cdot z = \frac{az+b}{cz+d} \tag{1.1}$$

for any $\begin{pmatrix} a & b \\ c & d \end{pmatrix} \in \mathrm{SL}_2(\mathbf{R})$ (see Figure 1.)

This means that, for any $z \in \mathbf{H}$, there exists some element $\gamma \in \mathrm{SL}_2(\mathbf{Z})$ such that $\gamma \cdot z \in F$, and in fact γ is usually unique (it is unique if $\gamma \cdot z$ is in the interior of F). Consider in particular a prime number p, and the p points

$$z_0 = \frac{i}{p},\ z_1 = \frac{1+i}{p}, \ldots, z_{p-1} = \frac{p-1+i}{p},$$

in $\mathbf{H}$. There are corresponding points $w_0, \ldots, w_{p-1}$ in F, each equivalent to the respective point z_j under the action of $\mathrm{SL}_2(\mathbf{Z})$. Where are these

Ph. M. was partially supported by the SNF (grant 200021-137488) and the ERC (Advanced Research Grant 228304). É. F. thanks ETH Zürich, EPF Lausanne and the Institut Universitaire de France for financial support.

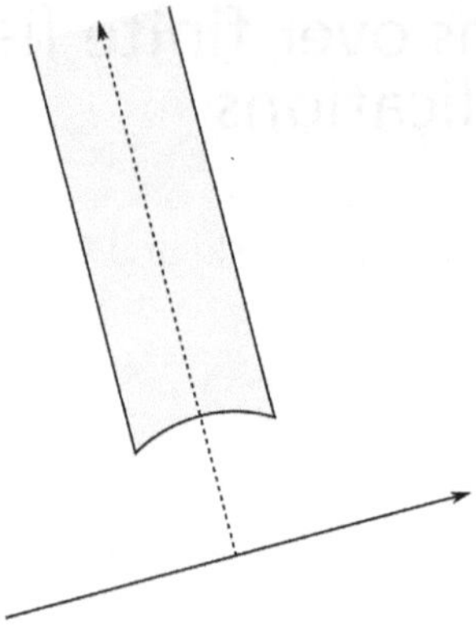

Figure 1. The leaning fundamental domain of Pisa.

points? Experiments quickly show that, as p increases, the points tend to range all over F. Indeed, one can prove that they become *equidistributed* as $p \to +\infty$, with respect to the probability measure

$$d\mu = \frac{3}{\pi}\frac{dxdy}{y^2}$$

on F (one checks indeed that $\mu(F) = 1$), which is a natural measure here because it is $\mathrm{SL}_2(\mathbf{R})$-invariant: $\mu(g^{-1}(A)) = \mu(A)$ for any $g \in \mathrm{SL}_2(\mathbf{R})$ and any measurable set $A \subset \mathbf{H}$. This equidistribution property means that

$$\lim_{p\to+\infty} \frac{1}{p}\sum_{j=0}^{p-1} \varphi\Big(\frac{j+w}{p}\Big) = \int_F \varphi(z)d\mu(z),$$

for any fixed $w \in \mathbf{H}$ (the case above being $w = i$) and for all continuous functions with compact support φ on $\mathbf{H}$ which are $\mathrm{SL}_2(\mathbf{Z})$-invariant: $\varphi(\gamma\cdot z) = \varphi(z)$ for all $z \in \mathbf{H}$.

In particular, in the spirit of Buffon's original needle problem, the following game is *fair*: for a large prime p, and a "random" integer $0 \leq j \leq p-1$, Players A and B take a bet as to whether the imaginary part of the corresponding w_j is $\geq 6/\pi$ or not. (Indeed, the measure of the subset

$$F_1 = \Big\{z \in F \mid \mathrm{Im}(z) \geq \frac{6}{\pi}\Big\} \subset F$$

is easily computed to be $1/2$; thus, more precisely, the game is only fair in the limit when $p \to +\infty$.)

One picturesque question which our work can address in this context, is the following: can one player gain an edge in this game by selecting her bet according to some algebraic property of j modulo p, for instance by determining if j is of the form $f(n)$ modulo p for some fixed polynomial $f \in \mathbf{Z}[X]$ and $n \in \mathbf{Z}/p\mathbf{Z}$, and selecting her bet based on this value? As we will explain, this is not possible, at least asymptotically if p tends to infinity.

The key to solving this problem turns out to involve an intricate combination of methods of analytic number theory and concepts and results of algebraic geometry over finite fields. More precisely, the crucial notion is that of *trace functions* over a finite field k, which are certain complex-valued functions $k \longrightarrow \mathbf{C}$ which have very strong algebraic features. Most importantly, and as the critical ingredient in most of our applications, these functions satisfy a form of quasi-orthogonality, that follows from a very general form of the Riemann Hypothesis for algebraic varieties over finite fields, due to Deligne [4].

In the motivating problem, these functions can be seen to arise naturally. Suppose $f \in \mathbf{Z}[X]$ is a fixed non-constant polynomial, and we wish to show that there is no gain in the game that may be derived by betting that the point w_j is in F_1 if and only if j is a value of f, *i.e.*, if and only if $j = f(n)$ for some $n \in \mathbf{Z}/p\mathbf{Z}$. If we denote by ξ_p the function defined for $0 \leq j \leq p-1$ as the characteristic function of the set $f(\mathbf{Z}/p\mathbf{Z})$ of values of f, and by φ_1 the characteristic function of F_1, this fairness property amounts to showing that

$$\frac{1}{p}\Big(\sum_{j=0}^{p-1}\xi_p(j)\varphi_1(w_j) + \sum_{j=0}^{p-1}(1-\xi_p(j))(1-\varphi_1(w_j))\Big) \longrightarrow \frac{1}{2}$$

as $p \to +\infty$ (since the left-hand side is the proportion of success when betting that $w_j \in F_1$ using the strategy we described). What we actually show is that if

$$\delta_p = \frac{1}{p}\sum_{j=0}^{p-1}\xi_p(j)$$

is the average value of ξ_p, then

$$\frac{1}{p}\sum_{j=0}^{p-1}(\xi_p(j)-\delta_p)\varphi_1(w_j) \longrightarrow 0,$$

$$\frac{1}{p}\sum_{j=0}^{p-1}((1-\xi_p(j))-(1-\delta_p))(1-\varphi_1(w_j)) \longrightarrow 0,$$

from which the limit of the sum above is

$$\lim_{p\to+\infty} \frac{1}{p}\sum_{j=0}^{p-1} \varphi_1(w_j),$$

which converges to $1/2$ because the set F_1 was chosen so that the original game is fair.

These ideas, and the techniques we developed, have many other applications than the one we have described above, and we will survey, sometimes briefly, most of our main results.

More generally, we prove (considering only the first limit above, the second being just a variant) that

$$\frac{1}{p}\sum_{j=0}^{p-1}(\xi_p(j)-\delta_p)\varphi\Bigl(\frac{j+i}{p}\Bigr) \longrightarrow 0 \tag{1.2}$$

for all continuous and compactly supported functions φ on $\mathbf{H}$. A standard limiting procedure then extends the result to the characteristic function of F_1. Now, according to the well-known Weyl criterion, we may limit our attention to certain well-chosen functions. There is a standard choice of such functions in our case, which consists of the square-integrable eigenfunctions φ of the hyperbolic Laplace operator, together with certain other functions which we will not discuss further (which arise due to the non-compactness of F). The eigenfunctions φ are analogues of the eigenfunctions of the standard Laplace operator $-d^2/dx^2$ for periodic functions on $\mathbf{R}/\mathbf{Z}$, namely the exponentials $x\mapsto e(hx)$ for $h\in\mathbf{Z}$, where we denote $e(z)=e^{2i\pi z}$. They have an expansion

$$\varphi(z) = \sum_{m\in\mathbf{Z}-\{0\}} \lambda_\varphi(m)W_\varphi(2\pi|m|y)e(mx),$$

where W_φ is a certain Whittaker-Bessel function (depending only on the eigenvalue of φ for the hyperbolic Laplace operator) and the Fourier coefficients $\lambda_\varphi(m)$ are complex numbers. Now observe that the left-hand side of (1.2) becomes

$$\sum_{m\neq 0}\lambda_\varphi(m)K_p(m)W_\varphi\Bigl(\frac{2\pi|m|}{p}\Bigr) \tag{1.3}$$

where

$$K_p(m) = \frac{1}{p}\sum_{j=0}^{p-1}(\xi_p(j)-\delta_p)e\Bigl(\frac{mj}{p}\Bigr). \tag{1.4}$$

This function K is naturally defined on $\mathbf{Z}/p\mathbf{Z}$, and its values on $\{0, ..., p-1\}$ are obtained by reduction modulo p. We consider it to be of algebraic nature because it is a discrete Fourier transform modulo p of the function $\xi_p - \delta_p$, which has an algebraic definition in terms of the polynomial f. We have in fact proved in [7] that the limiting formula

$$\frac{1}{p}\sum_{m\neq 0}\lambda_\varphi(m)K_p(m)V(m) \longrightarrow 0 \tag{1.5}$$

holds as $p \longrightarrow +\infty$, for much more general cases of *trace functions* K_p and more general "nice" weight functions $V(m)$ than $V(m) = W_\varphi(2\pi|m|/p)$. This solves the fairness question we used as a motivation, and in fact it is a much more general and widely applicable result.

The outline of the remainder of this survey is the following. In the next section, we define rigorously trace functions; there appears then a crucial definition for analytic applications, that of the *conductor* of a trace function, which measures its complexity, in such a way that uniformity with respect to p may be considered. These definitions are illustrated in Section 3 with many examples. We next discuss the crucial, extremely deep and extremely powerful quasi-orthogonality property, which follows from the Riemann Hypothesis over finite fields, and how we use it in [7] and [8]. The last section discusses another, very concrete, application to the distribution of certain arithmetic functions in arithmetic progressions to large moduli, following [9].

We do not discuss some other papers, contenting ourselves with the following short indications: (1) in [10], we show that trace functions are "Gowers-uniform to all order", unless they have a very special shape, providing in particular the first explicit examples of functions on $\mathbf{Z}/p\mathbf{Z}$ with Gowers norms as small as those of "random" functions; (2) in [11], we use ideas of spherical codes to (roughly) bound from above the number of trace functions modulo p with bounded conductor; (3) in [12], we introduce a new method to estimate short exponential sums modulo primes, of length very close to $\sqrt{p}$, and obtain improvements for trace functions of the classical Polyá-Vinogradov bound.

ACKNOWLEDGEMENTS. We warmly thank U. Zannier for inviting one of us to present these results as a De Giorgi Colloquium at the Scuola Normale Superiore di Pisa and giving us the opportunity to present a written account of our work. We also thank F. Jouve and R. Cluckers for comments on this text.

1.1. Notation

We recall here some basic notation.
– The letters p will always refer to a prime number; for a prime p, we write $\mathbf{F}_p$ for the finite field $\mathbf{Z}/p\mathbf{Z}$. For a set X, $|X|$ is its cardinality, a non-negative integer or $+\infty$.
– The Landau and Vinogradov notation $f = O(g)$ and $f \ll g$ are synonymous, and $f(x) = O(g(x))$ for all $x \in D$ means that there exists an "implied" constant $C \geq 0$ (which may be a function of other parameters) such that $|f(x)| \leq Cg(x)$ for all $x \in D$. This definition *differs* from that of N. Bourbaki [2, Chap. V] since the latter is of topological nature. We write $f \asymp g$ if $f \ll g$ and $g \ll f$. On the other hand, the notation $f(x) \sim g(x)$ and $f = o(g)$ are used with the asymptotic meaning of loc. cit.

2. Trace functions: definition

We present in this section the definition of trace functions over a finite field $\mathbf{F}_p$, and of the invariant which measures their complexity. This definition can, in fact, be presented from three different points of view: using automorphic forms (over $\mathbf{F}_p(T)$), using Galois representations (of the Galois group of $\mathbf{F}_p(T)$), or using étale sheaf theory. We use here the second because it is the most elementary, but the last is in fact the most convenient in many respects because it leads to the most flexible formalism, as we will describe. In order to be consistent with the terminology of our papers, we will use the language of sheaves after this section.

It is at least equally important to know the most common examples of trace functions, and for many applications to analytic number theory, one can in fact view trace functions as a kind of black box, building on the known very concrete examples and on the formalism these functions satisfy, especially the deep quasi-orthogonality property that encapsulates the Riemann Hypothesis over finite fields, as discussed in Section 4. In particular, readers who find the following definitions rather too abstract can just go through them very quickly, and concentrate their attention on the examples in the next section.

Let p be a fixed prime. We must first fix a different prime $\ell \neq p$, which plays an auxiliary role, and fix an isomorphism (of fields)

$$\iota : \bar{\mathbf{Q}}_\ell \longrightarrow \mathbf{C}$$

between a given algebraic closure of the field of ℓ-adic numbers and the field of complex numbers. In fact, we will mostly view this isomorphism as an algebraic identification, so that the reader may view $\bar{\mathbf{Q}}_\ell$ as just another name for $\mathbf{C}$; the main difference between the two, which is very important for the theory, is their different topological nature.

Let $K = \mathbf{F}_p(T)$ be the field of rational functions with coefficients in $\mathbf{F}_p$, and let $\bar{K}$ denote a separable closure of K (in which an algebraic closure $\bar{\mathbf{F}}_p$ of $\mathbf{F}_p$ is contained); elements of $\bar{K}$ can therefore be interpreted as "algebraic functions" on the projective line $\mathbf{P}^1(\bar{\mathbf{F}}_p)$, such as $\sqrt{f(X)}$ where $f \in \mathbf{F}_p[X]$ is a polynomial.

We let Π denote the Galois group of $\bar{K}$ over K. This groups contains a normal subgroup Π^g defined as the Galois group of $\bar{K}$ over the subfield $\tilde{K} = \bar{\mathbf{F}}_p(T)$, and the quotient Π/Π^g is naturally isomorphic to the Galois group of $\bar{\mathbf{F}}_p$ over $\mathbf{F}_p$, which is (topologically) generated by the arithmetic Frobenius automorphism $x \mapsto x^p$, or by its inverse, which is called the geometric Frobenius automorphism.

Definition 2.1 (ℓ-adic representation). An *ℓ-adic Galois representation ρ over* $\mathbf{F}_p$ is a continuous group homomorphism

$$\rho : \Pi \longrightarrow \mathrm{GL}(V),$$

for some finite-dimensional $\bar{\mathbf{Q}}_\ell$-vector space V, such that, for all but finitely many $x \in \mathbf{P}^1(\bar{\mathbf{F}}_p)$, the inertia group I_x at x is contained in the kernel of ρ.

The dimension of V is called the rank of ρ, and the set of $x \in \mathbf{P}^1(\bar{\mathbf{F}}_p)$ where I_x does not act trivially is called the set of singularities, or the set of ramification points, of ρ, and is denoted $\mathrm{Sing}(\rho)$. One also says that ρ is *lisse at* $x \in \mathbf{P}^1(\bar{\mathbf{F}}_p)$ if I_x acts trivially on V.

We also say that ρ is a *middle-extension ℓ-adic sheaf* modulo p, or sometimes just *ℓ-adic sheaf*. When using this language, we usually use curly letters, such as $\mathcal{F}$, instead of ρ.

In this definition, as in classical algebraic number theory, the inertia group I_x is defined as the normal subgroup of the decomposition group

$$D_x = \{\gamma \in \Pi \mid \gamma(f(x)) = 0 \text{ for all } f \in \bar{K} \text{ such that } f(x) = 0\}$$

characterized by the condition

$$I_x = \{\gamma \in D_x \mid \gamma(f(x)) = f(x) \text{ for all } f \in \bar{K}\}.$$

These definitions make sense, because we can view an algebraic function $f \in \bar{K}$ as a "multi-valued function" on $\mathbf{P}^1(\bar{\mathbf{F}}_p)$: although $f(x)$ is not uniquely defined, all possible values are conjugates under the Galois group of $\bar{\mathbf{F}}_p/\mathbf{F}_p$, and the conditions defining D_x and I_x are invariant under this action (we also use the fact that γ also acts on $\bar{\mathbf{F}}_p \subset \bar{K}$).

It is immediately clear that this definition gives a relatively flexible formalism: we can form direct sums $\rho_1 \oplus \rho_2$, tensor products $\rho_1 \otimes \rho_2$, dual $\mathrm{D}(\rho)$, of Galois representations, and we can define subrepresentations, quotient representations, and morphisms of representations (and therefore we can speak of isomorphic representations, or equivalently of isomorphic middle-extension sheaves).

For us, the point of this definition is that to each ℓ-adic representation ρ is naturally attached a function $\mathbf{F}_p \longrightarrow \mathbf{C}$, which is called its trace function. From the representation theory point of view, it is just the restriction of the character $\operatorname{Tr}\rho$ of the representation to special (conjugacy classes of) elements in Π.

Definition 2.2 (Trace function). Let ρ be an ℓ-adic Galois representation over $\mathbf{F}_p$. The *trace function* of ρ is the function

$$t_\rho : \mathbf{F}_p \longrightarrow \mathbf{C}$$

defined by

$$t_\rho(x) = \iota\Big(\operatorname{Tr}(\rho(\sigma_x \mid V^{I_x}))\Big),$$

for $x \in \mathbf{F}_p$, where σ_x denotes the conjugacy class of the geometric Frobenius automorphism at x, which generates topologically the quotient $D_x/I_x \simeq \mathrm{Gal}(\bar{\mathbf{F}}_p/\mathbf{F}_p)$ and V^{I_x} denotes the subspace of V invariant under the action of I_x, on which D_x/I_x acts naturally, while $\operatorname{Tr}(g \mid W)$ denotes the trace of an endomorphism g acting on a vector space W.

We have

$$t_{\rho_1\oplus\rho_2} = t_{\rho_1} + t_{\rho_2}$$

and

$$t_{\rho_1\otimes\rho_2}(x) = t_{\rho_1}(x)t_{\rho_2}(x)$$

for all x such that I_x acts trivially on ρ_1 and ρ_2 (at least).

In particular, the set of trace functions of ℓ-adic representations modulo p is closed under addition, and "almost" closed under products.

This set is infinite, and although not every function is of this form, it is dense for the uniform norm on functions modulo p. This implies that very few interesting analytic statements can be expected to hold for *all* trace functions. However, in applications, the trace functions that arise have two extra properties which rigidify the situation. Together, they explain the versatility and power of trace functions in analytic number theory.

The first condition is a restriction on the eigenvalues of the action of the Frobenius automorphisms on V.

Definition 2.3 (Weight 0 representation). An ℓ-adic representation ρ modulo p, acting on V, is *pointwise of weight* 0 if and only if the following condition holds:

For all finite extensions $k/\mathbf{F}_p$ and all $x \in \mathbf{P}^1(k)$ such that I_x acts trivially on V, the eigenvalues of $\rho(\sigma_x)$ are algebraic numbers α such that all Galois conjugates of α have modulus 1.

In fact, it follows from non-trivial results of Deligne that one need only check the condition for x in the complement U of a finite set of points of $\mathbf{P}^1(\bar{\mathbf{F}}_p)$ such that ρ is lisse at every point in U, and also that if ρ is not lisse at $x \in \mathbf{P}^1(k)$ for some finite extension $k/\mathbf{F}_p$, it nevertheless satisfies the condition that there exists an integer $w \leq 0$ such that the eigenvalues of $\rho(\sigma_x)$ are algebraic numbers α such that all Galois conjugates of α have modulus $|k|^w$ (in particular, have modulus ≤ 1).

Note that if ρ, ρ_1 and ρ_2 are of weight 0, then so are $\rho_1 \oplus \rho_2$ and $\rho_1 \otimes \rho_2$, and the dual of ρ, as well as any subrepresentation or quotient representation of ρ. Furthermore, one can show that the trace function of the dual of a weight zero representation ρ is the complex conjugate of the trace function of ρ (which is easy at all unramified points, and the point is that it is also true for the possible ramified points, by a result of Gabber.)

A simple and immediate consequence of the definition is that the trace function of a representation of weight 0 satisfies

$$|t_\rho(x)| \leq \dim(V)$$

for all $x \in \mathbf{F}_p$. This is a first indication of how one can control the complexity of the trace function of an ℓ-adic representation.

Definition 2.4 (Trace function modulo p). Let p be a prime number. A *trace function modulo p* is a function $t : \mathbf{F}_p \longrightarrow \mathbf{C}$ such that $t = t_{\mathcal{F}}$ for some ℓ-adic middle-extension sheaf $\mathcal{F}$ of weight 0.

Note that (as we will clearly see below) the sheaf $\mathcal{F}$ is not unique. However, it is unique if one restricts one's attention to representations with small complexity, in the sense that the conductor, which we now define, is small enough compared with p. This definition is absolutely essential for all analytic applications.

Definition 2.5 (Conductor). Let $\rho : \Pi \longrightarrow \mathrm{GL}(V)$ be an ℓ-adic representation modulo p of weight 0. The *conductor* of ρ is

$$\mathbf{c}(\rho) = \dim(V) + |\operatorname{Sing}(\rho)| + \sum_{x \in \operatorname{Sing}(\rho)} \operatorname{Swan}_x(\rho),$$

where for a ramified point $x \in \mathbf{P}^1(\bar{\mathbf{F}}_p)$, we denote by $\operatorname{Swan}_x(\rho)$ the Swan conductor of ρ at x.

The precise definition of the Swan conductor, which measures fine properties of the representation ρ at a ramified point, can be found for instance in [19, Ch. 1]. It is a rather subtle invariant, and we will mostly attempt to illustrate its meaning and properties with examples. For the moment, we will only mention that $\mathrm{Swan}_x(\rho)$ is a non-negative integer. When $\mathrm{Swan}_x(\rho) = 0$, one says that ρ is *tamely ramified* at x. In a number of important cases, ρ is tamely ramified at all $x \in \mathrm{Sing}(\rho)$, in which case one says that ρ itself is *tamely ramified*.

Finally, we remark that this definition of trace functions is not sufficient for certain constructions and applications, where slightly more general objects (constructible ℓ-adic sheaves) appear more naturally (see for instance [13]).

3. Trace functions: examples

The examples in this section are not only concrete examples of functions modulo p which are trace functions, but also examples of *operations* which may be performed on trace functions and lead to other trace functions. In all cases, it is very important to understand at least an upper-bound for the conductor of the associated ℓ-adic sheaves (or representations).

3.1. Characters

The simplest examples of trace functions are *character values*, involving either additive or multiplicative characters, or a product of them. Specifically, let p be a prime, and let

$$\psi \,:\, \mathbf{F}_p \longrightarrow \mathbf{C}^\times$$

be a non-trivial additive character (for instance, $\psi(x) = e(x/p)$ for $x \in \mathbf{F}_p$). Fix a rational function $f \in \mathbf{F}_p(X)$, which has no pole of order divisible by p, and define

$$t(x) = \begin{cases} e\left(\dfrac{f(x)}{p}\right) & \text{if } x \text{ is not a pole of } f \\ 0 & \text{otherwise.} \end{cases}$$

Then one can show that there exists an ℓ-adic middle-extension sheaf modulo p, denoted $\mathcal{L}_{\psi(f)}$, such that t is the trace function of $\mathcal{L}_{\psi(f)}$. Indeed, in contrast with most other examples we will discuss, this construction is elementary (see, *e.g.*, [18, §11.11] for a discussion). A sheaf of the form $\mathcal{L}_{\psi(f)}$ is called an *Artin-Schreier sheaf*.

This sheaf is of rank 1 and ramified precisely at the poles of f (this applies also to the possible ramification at ∞, and uses the assumption that no pole is of order divisible by p). It has weight 0 (the image under ι of the only eigenvalue of Frobenius at any unramified $x \in \mathbf{P}^1(\mathbf{F}_p)$ is $\psi(f(x))$). As for the Swan conductors, if $x \in \mathbf{P}^1(\bar{\mathbf{F}}_p)$ is a pole of f, then one shows that $\mathrm{Swan}_x(\mathcal{L}_{\psi(f)})$ is *at most* equal to the order of the pole at x, and that there is equality at least if the pole is of order $< p$. Hence, if $f_2 \in \mathbf{F}_p[X]$ is the denominator of f, we have

$$\mathbf{c}(\mathcal{L}_{\psi(f)}) \leq 1 + 2\deg(f_2).$$

In particular, suppose we take now a rational function $f \in \mathbf{Q}(X)$, and we write $f = f_1/f_2$ with $f_i \in \mathbf{Z}[X]$. For all primes p large enough, we may reduce f modulo p to consider $f_1/f_2 \in \mathbf{F}_p(X)$. For each such p, we can form the corresponding sheaf $\mathcal{L}_{\psi(f)}$ modulo p, and what is essential is that the conductor of these sheaves is bounded by a constant depending only on f, and *not* on the prime p.

Similarly, let $\chi : \mathbf{F}_p^\times \longrightarrow \mathbf{C}^\times$ be a non-trivial multiplicative character, which can be seen as a Dirichlet character modulo p. Let $d \geq 2$ be the order of χ. Fix a rational function $f \in \mathbf{F}_p(X)$ such that f has no pole or zero of order divisible by d, and define

$$t(x) = \begin{cases} \chi(f(x)) & \text{if } x \text{ is not a zero or pole of } f \\ 0 & \text{otherwise.} \end{cases}$$

Then one can show (again, elementarily) that there exists an ℓ-adic middle-extension sheaf modulo p, denoted $\mathcal{L}_{\chi(f)}$, such that t is equal to the trace function of $\mathcal{L}_{\chi(f)}$ (through ι). Such sheaves are called *Kummer sheaves*.

This representation is of rank 1 and ramified precisely at the zeros and poles of f (in $\mathbf{P}^1(\bar{\mathbf{F}}_p)$). It has weight 0, the only (image under ι of an) eigenvalue of Frobenius at any unramified $x \in \mathbf{P}^1(\mathbf{F}_p)$ being equal to $\chi(f(x))$. Furthermore, it is *tamely ramified*. Thus

$$\mathbf{c}(\mathcal{L}_{\chi(f)}) \leq 1 + \deg(f_1) + \deg(f_2)$$

if $f = f_1/f_2$ with $f_i \in \mathbf{F}_p[X]$ coprime. As before, if we obtain f by reduction modulo p from a fixed rational function with rational coefficients, this family of sheaves indexed by p has conductor bounded independently of p.

One can combine these examples by tensor product, which amounts to multiplying the trace functions: if $\chi_1, \ldots, \chi_r$ are finitely many distinct non-trivial multiplicative characters, and $f_1, \ldots, f_{r+1}$ are rational functions in $\mathbf{F}_p(X)$, the middle-extension sheaf

$$\rho = \mathcal{L}_{\chi_1(f_1)} \otimes \cdots \otimes \mathcal{L}_{\chi_r(f_r)} \otimes \mathcal{L}_{\psi(f_{r+1})}$$

has rank 1, weight 0, and it is unramified[1] at least outside of the union of the poles of f_{r+1} and the zeros and poles of $f_1, \ldots, f_r$, it has trace function

$$t_{\rho(x)} = \chi_1(f_1(x)) \cdots \chi_r(f_r(x)) e\Big(\frac{f_{r+1}(x)}{p}\Big)$$

for all unramified $x \in \mathbf{F}_p$, and has conductor

$$\mathbf{c}(\rho) \leq 1 + \sum_{i=1}^{r} (\deg(g_i) + \deg(h_i)) + \deg(h_{r+1}),$$

where $f_i = g_i/h_i$ with $g_i, h_i \in \mathbf{F}_p[X]$ coprime.

Example 3.1. If we take $f(X) = X^p - X \in \mathbf{F}_p[X]$, the trace function associated to $\mathcal{L}_{\chi(f)}$, for χ non-trivial, satisfies

$$t_{\mathcal{L}_{\chi(f)}}(x) = \chi(x^p - x) = 0$$

for all $x \in \mathbf{F}_p$. Thus, for any trace function t associated to a sheaf $\mathcal{F}$, we also have

$$t = t_{\mathcal{F} \oplus \mathcal{L}_{\chi(f)}},$$

illustrating the non-uniqueness of $\mathcal{F}$. Note however that this second sheaf has huge conductor in terms of p:

$$\mathbf{c}(\mathcal{F} \oplus \mathcal{L}_{\chi(f)}) \geq p.$$

3.2. Point-counting functions

The second class of examples is also relatively elementary. We consider a non-constant squarefree polynomial $f \in \mathbf{F}_p[X]$, and for $x \in \mathbf{F}_p$, we denote

$$n_f(x) = |\{y \in \mathbf{F}_p \mid f(y) = x\}|,$$

the number of pre-images of x in the field $\mathbf{F}_p$. In particular, $n_f(x) = 0$ if x is not of the form $x = f(y)$ for *some* $y \in \mathbf{F}_p$. One can then construct an ℓ-adic middle-extension sheaf $\mathcal{F}_f$ such that $t_{\mathcal{F}_f}(x) = n_f(x)$ for all $x \in \mathbf{F}_p$. (Indeed, if $K = \mathbf{F}_p(f(X))$, we have a Galois extension $\mathbf{F}_p(X)/\mathbf{F}_p(f(X))$, and since $\mathbf{F}_p(f(X))$ is isomorphic to $\mathbf{F}_p(X)$, this

[1] Depending on the χ_i and f_i, it might be unramified at some extra points.

gives a homomorphism $\mathrm{Gal}(\bar{K}/K) \longrightarrow \mathrm{Gal}(\bar{K}/K)$ with image Π_f of finite index in $\Pi = \mathrm{Gal}(\bar{K}/K)$; then the representation ρ_f which "is" $\mathcal{F}_f$ can be defined as the induced representation $\mathrm{Ind}_{\Pi_f}^{\Pi}(\bar{\mathbf{Q}}_\ell)$.)

Because the induced representation always contains a copy $\mathbf{1}$ of the trivial representation, it is often convenient to consider $\mathcal{F}'_f = \mathcal{F}_f/\mathbf{1}$, which has trace function

$$n_f^0(x) = t_{\mathcal{F}'_f}(x) = n_f(x) - 1.$$

The point (for applications) is that 1 is the average value of $n_f(x)$, so we have

$$\sum_{x\in\mathbf{F}_p} n_f^0(x) = \sum_{x\in\mathbf{F}_p} n_f(x) - p = \sum_{x\in\mathbf{F}_p}\sum_{\substack{y\in\mathbf{F}_p\\ f(y)=x}} 1 - p = \sum_{y\in\mathbf{F}_p} 1 - p = 0.$$

The representation $\mathcal{F}'_f$ has rank $\deg(f) - 1$. It has weight 0, and it is unramified at all $x \in \mathbf{P}^1(\bar{\mathbf{F}}_p)$ such that the pre-image $f^{-1}(x) \subset \mathbf{P}^1(\bar{\mathbf{F}}_p)$ (with coefficients in $\bar{\mathbf{F}}_p$ now) consists of $\deg(f)$ different points, or in other words, at all regular values of f (where a regular value is one which is not a critical value, and a critical value is the image of a critical point $y \in \bar{\mathbf{F}}_p$, *i.e.*, of a root of f').

If $p > \deg(f)$, it is known that $\mathcal{F}_f$ and $\mathcal{F}'_f$ are tamely ramified. In particular, for $\deg(f) \geq 2$ (so that $\mathcal{F}'_f$ is non-zero), we have

$$\mathbf{c}(\mathcal{F}'_f) \leq \deg(f) - 1 + \deg(f) - 1 = 2\deg(f) - 2.$$

The trace function of $\mathcal{F}_f$ counts the number of solutions in $\mathbf{F}_p$ of the equation $f(y) = x$, and therefore it is supported on the subset $f(\mathbf{F}_p) \subset \mathbf{F}_p$. In our motivating problem in Section 1, on the other hand, we considered the function ξ_p which is the characteristic function of this set. This is also related to trace functions of ℓ-adic sheaves, although it is not quite one. Rather, one shows (the construction is again elementary, see [8, Prop. 6.7]) that if $\deg(f) > p$, there exist middle-extension sheaves $\mathcal{F}_{f,i}$, $1 \leq i \leq m \leq \deg(f)$, all geometrically non-trivial, pointwise pure of weight 0 and tamely ramified, and algebraic numbers c_0, c_i for $1 \leq i \leq m$, such that

$$\xi_p(x) = c_0 + \sum_{i=1}^{m} c_i t_{\mathcal{F}_{f,i}}(x) \tag{3.1}$$

for all $x \in \mathbf{F}_p - S$, where $S \subset \mathbf{F}_p$ is a finite set of cardinality $\leq \deg(f)$. Moreover, the parameters m, $|c_i|$, $\mathbf{c}(\mathcal{F}_{f,i})$ which measure the complexity of this decomposition are all bounded in terms of $\deg(f)$, and

$$c_0 = \frac{|f(\mathbf{F}_p)|}{p} + O(p^{-1/2}),$$

where the implied constant depends only on $\deg(f)$.

Remark 3.2. If $f = X^2 + a$, then (for all odd primes p) the trace functions $n_f(x)$ and $n_f^0(x)$ are given by

$$n_f(x) = 1 + \Big(\frac{x-a}{p}\Big), \qquad n_f^0(x) = \Big(\frac{x-a}{p}\Big),$$

where $\big(\frac{\cdot}{p}\big)$ is the Legendre symbol, while ξ_p is given by

$$\xi_p(x) = \frac{1}{2}\Big(1 + \Big(\frac{x-a}{p}\Big)\Big),$$

for $x \neq a$, and $\xi_p(a) = 1$. Thus we have

$$\mathcal{F}_f = \mathbf{1} \oplus \mathcal{L}_{\left(\frac{X-a}{p}\right)}, \qquad \mathcal{F}_f' = \mathcal{L}_{\left(\frac{X-a}{p}\right)}.$$

The general decomposition is therefore a (non-obvious) generalization of this fact.

3.3. Exponential sums

Our next example may seem very specialized, but it plays a critical role in many deep results in analytic number theory. Let ψ be the non-trivial additive character ψ modulo p given by $\psi(x) = e(x/p)$ (note that this depends on p). For an integer $n \geq 1$, we consider the (normalized) *hyper-Kloosterman sums* $\mathrm{Kl}_n(x;p)$ defined by

$$\mathrm{Kl}_n(x;p) = \frac{(-1)^{n-1}}{p^{(n-1)/2}} \sum_{\substack{y_1,\ldots,y_n \in \mathbf{F}_p^\times \\ y_1\cdots y_n = x}} \!\!\!\cdots \sum \psi(y_1 + \cdots + y_n)$$

for $x \in \mathbf{F}_p^\times$. For instance we have $\mathrm{Kl}_1(x;p) = e(x/p)$

$$\mathrm{Kl}_2(x;p) = -\frac{1}{\sqrt{p}} \sum_{y \in \mathbf{F}_p^\times} e\Big(\frac{xy + \bar{y}}{p}\Big),$$

which is the classical Kloosterman sum with parameter x. It is now a highly non-trivial fact that there exists an ℓ-adic middle-extension sheaf $\mathcal{K}\ell_n$ modulo p, called a *Kloosterman sheaf* such that

$$t_{\mathcal{K}\ell_n}(x) = \mathrm{Kl}_n(x;p)$$

for $x \in \mathbf{F}_p^\times$. In fact, as far as we are aware, there is no proof of the existence of this representation directly in the framework of Galois representations: one must construct it as an ℓ-adic sheaf (a construction due to Deligne, and extensively studied by Katz [19], which was recently generalized by Heinloth, Ngô and Yun [17]).

Among the results of Deligne and Katz concerning Kloosterman sheaves are the following: $\mathcal{K}\ell_n$ is geometrically irreducible, it is of rank n, pointwise pure of weight 0, ramified only at 0 (if $n \geq 2$) and ∞ in $\mathbf{P}^1(\bar{\mathbf{F}}_p)$, and the ramification is tame at 0 (*i.e.*, $\mathrm{Swan}_0(\mathcal{K}\ell_n) = 0$) and wild at ∞ with $\mathrm{Swan}_\infty(\mathcal{K}\ell_n) = 1$. Thus, for every prime p and $n \geq 2$, we have

$$\mathbf{c}(\mathcal{K}\ell_n) = n + 2 + 1 = n + 3,$$

and it is again crucial that the conductor is bounded independently of p.

To see how deep such results are, note that since $\mathcal{K}\ell_n$ is of weight 0 and unramified at $x \in \mathbf{F}_p^\times$, it follows that

$$|\,\mathrm{Kl}_n(x; p)| \leq n$$

for all primes p and all $x \in \mathbf{F}_p^\times$, for instance

$$\Big|\sum_{y \in \mathbf{F}_p^\times} e\Big(\frac{xy + \bar{y}}{p}\Big)\Big| \leq 2\sqrt{p}$$

for $x \in \mathbf{F}_p^\times$. This bound is the well-known *Weil bound for Kloosterman sums*, and it has countless applications in analytic number theory (due particularly to the presence of Kloosterman sums in the theory of automorphic forms, see *e.g.* [21] for a survey).

3.4. Operating on trace functions

It is a fundamental aspect of ℓ-adic sheaves and their trace functions that a flexible formalism is available in their study, and for applications. Besides the standard operations mentioned above (direct sums, tensor product, dual), we will illustrate this point here with one particular operation that is very relevant for our papers, in particular in [7–9]: the Fourier transform.

For a prime number p, a non-trivial additive character ψ and a function $\varphi : \mathbf{F}_p \longrightarrow \mathbf{C}$, we define here the Fourier transform $\mathrm{FT}_\psi(\varphi) : \mathbf{F}_p \longrightarrow \mathbf{C}$ by the formula

$$\mathrm{FT}_\psi(\varphi)(t) = -\frac{1}{\sqrt{p}} \sum_{x \in \mathbf{F}_p} \varphi(x)\psi(xt)$$

for $t \in \mathbf{F}_p$. If φ is a trace function as we defined them, the Fourier transform can not always be one, because the Fourier transform of a constant, for instance, is a delta function, which does not fit our framework well. However, exploiting the deep fact (due to Deligne [4, (3.4.1)]) that a middle-extension sheaf modulo p of weight 0 is, geometrically (*i.e.*, over $\bar{\mathbf{F}}_p$) a direct sum of irreducible sheaves over $\bar{\mathbf{F}}_p$, one can define a *Fourier sheaf* modulo p to be one where no such geometrically irreducible component is isomorphic to an Artin-Schreier sheaf $\mathcal{L}_{\psi(aX)}$ for some $a \in \bar{\mathbf{F}}_p$. Then Deligne showed that there exists an operation

$$\mathcal{F} \mapsto \mathrm{FT}_{\psi}(\mathcal{F})$$

at the level of Fourier ℓ-adic sheaves with the property that

$$t_{\mathrm{FT}_{\psi}(\mathcal{F})} = \mathrm{FT}(t_{\mathcal{F}}),$$

i.e., the trace function of the Fourier transform of a sheaf $\mathcal{F}$ is equal to the Fourier transform of the trace function of $\mathcal{F}$. This operation was studied in depth by Laumon [22], Brylinski and Katz [19,20], and shown to satisfy the following properties (many of which are, intuitively, analogues of classical properties of the Fourier transform):

(1) If a Fourier sheaf $\mathcal{F}$ is pointwise of weight 0, then so is $\mathrm{FT}_{\psi}(\mathcal{F})$: this fact is extremely deep, as it relies on a refined application of the Riemann Hypothesis over finite fields.

(2) If $\mathcal{F}$ is geometrically irreducible, then so is $\mathrm{FT}_{\psi}(\mathcal{F})$ (as we will see in Section 4, this is to some extent an analogue of the unitarity property

$$\| \mathrm{FT}(\varphi))\|^2 = \|\varphi\|^2$$

of the Fourier transform of a function $\varphi : \mathbf{F}_p \longrightarrow \mathbf{C}$, where

$$\|\varphi\|^2 = \frac{1}{p} \sum_{x \in \mathbf{F}_p} |\varphi(x)|^2$$

is the standard L^2-norm.)

(3) Laumon [22] developed in particular a theory of "local Fourier transforms" which is an analogue of the stationary phase method in classical analysis, and which leads to very detailed information concerning the ramification properties of $\mathrm{FT}_{\psi}(\mathcal{F})$. Using this, we proved in [7] that the Fourier transform of sheaves has the important property that the conductor of $\mathrm{FT}_{\psi}(\mathcal{F})$ can be estimated solely in terms of the conductor of $\mathcal{F}$, and more precisely we showed:

$$\mathbf{c}(\mathrm{FT}_{\psi}(\mathcal{F})) \leq 10\mathbf{c}(\mathcal{F})^2.$$

This estimate is essential in analytic applications, since it implies that if p varies but $\mathcal{F}$ has bounded conductor, so do the Fourier transforms. In [13], we have extended such estimates to other linear transformations $\varphi \mapsto T\varphi$ of the type

$$(T\varphi)(x) = -\frac{1}{\sqrt{p}} \sum_{y \in \mathbf{F}_p} \varphi(y)\psi(f(x, y))$$

for arbitrary rational functions f.

Example 3.3. (1) As a first example, note that the function K_p defined by (1.4) for a fixed polynomial $f \in \mathbf{Z}[X]$ and ξ_p the characteristic function of $f(\mathbf{F}_p) \subset \mathbf{F}_p$ is (up to a factor $p^{1/2}$) the Fourier transform of $\xi_p - \delta_p$. By (3.1) and the remarks following, we see that we have (essentially)

$$K_p = \frac{1}{\sqrt{p}} \sum_{i=1}^{m} c_i t_{\mathrm{FT}(\mathcal{F}_{f,i})}, \tag{3.2}$$

i.e., $\sqrt{p}K_p$ is, if not a trace function, then at least a "short" linear combination of trace functions with bounded complexity in terms of $\deg(f)$. This is a crucial step in the proof of the results we mentioned in this first section.

(2) Consider the Artin-Schreier sheaf $\mathcal{L}_{\psi(X^{-1})}$ as in Section 3.1. Then we see that the Fourier transform $\mathcal{F} = \mathrm{FT}(\mathcal{L}_{\psi(X^{-1})})$ has trace function

$$t_{\mathcal{F}}(x) = -\frac{1}{\sqrt{p}} \sum_{y \in \mathbf{F}_p} \psi(y^{-1})\psi(xy) = \mathrm{Kl}_2(x; p).$$

In fact, this sheaf $\mathcal{F}$ is the same as the Kloosterman sheaf $\mathcal{K}\ell_2$ discussed in Section 3.3, and one can deduce all the basic properties of the latter from the general theory of the Fourier transform. The other Kloosterman sheaves $\mathcal{K}\ell_n$, for $n \geq 3$, can be constructed similarly using the operation of *multiplicative convolution* on trace functions.

4. Quasi-orthogonality of trace functions

The most important analytic property of trace functions lies in the quasi-orthogonality of trace functions of geometrically irreducible sheaves of weight 0, which is a very important and general form of the Riemann Hypothesis over finite fields as proved by Deligne [4].

Theorem 4.1 (Deligne). *Let p be a prime number, $\ell \neq p$ a prime distinct from p and let $\mathcal{F}_1$, $\mathcal{F}_2$ be geometrically irreducible ℓ-adic sheaves modulo p which are pointwise of weight* 0.

(1) *If $\mathcal{F}_1$ is not geometrically isomorphic to $\mathcal{F}_2$, then we have*

$$\Big|\sum_{x\in\mathbf{F}_p} t_{\mathcal{F}_1}(x)\overline{t_{\mathcal{F}_2}(x)}\Big| \leq 3\mathbf{c}(\mathcal{F}_1)^2\mathbf{c}(\mathcal{F}_2)^2 p^{1/2}. \tag{4.1}$$

(2) *If $\mathcal{F}_1$ is geometrically isomorphic to $\mathcal{F}_2$, then there exists a complex number α with modulus 1 such that*

$$t_{\mathcal{F}_1}(x) = \alpha t_{\mathcal{F}_2}(x)$$

for all $x \in \mathbf{F}_p$, and

$$\Big|\sum_{x\in\mathbf{F}_p} t_{\mathcal{F}_1}(x)\overline{t_{\mathcal{F}_2}(x)} - \alpha p\Big| \leq 3\mathbf{c}(\mathcal{F}_1)^2\mathbf{c}(\mathcal{F}_2)^2 p^{1/2}. \tag{4.2}$$

Note that, in that case, we have $\mathbf{c}(\mathcal{F}_1) = \mathbf{c}(\mathcal{F}_2)$.

To be precise, it follows from the Grothendieck-Lefschetz trace formula (see, *e.g.*, [3, Rapport, Th. 3.2]) and the Riemann Hypothesis [4, Th. 3.3.1] that both inequalities hold with right-hand side replaced by

$$(\dim H_c^1(\mathbf{A}^1 \times \bar{\mathbf{F}}_p, \mathcal{F}_1 \otimes \mathrm{D}(\mathcal{F}_2)))p^{1/2},$$

where we recall that $\mathrm{D}(\mathcal{F}_2)$ denotes the dual of $\mathcal{F}_2$, and from the Grothendieck-Ogg-Shafarevich formula for the Euler-Poincaré characteristic of a sheaf (see, *e.g.*, [19, 2.3.1]), one obtains relatively easily a bound

$$\dim H_c^1(\mathbf{A}^1 \times \bar{\mathbf{F}}_p, \mathcal{F}_1 \otimes \mathrm{D}(\mathcal{F}_2)) \leq 3\mathbf{c}(\mathcal{F}_1)^2\mathbf{c}(\mathcal{F}_2)^2$$

(see, *e.g.*, [11, Lemma 3.3]).

Remark 4.2. One useful interpretation of this result is as an approximate version of the orthogonality relations of characters of representations of finite (or compact) groups, which algebraically is related to Schur's Lemma. In particular, note that it implies that if $\mathbf{c}(\mathcal{F}_1)$ and $\mathbf{c}(\mathcal{F}_2)$ are small enough (roughty $\ll p^{1/8}$ in this version), the condition $t_{\mathcal{F}_1} = t_{\mathcal{F}_2}$ of equality of the trace functions suffices to imply that $\mathcal{F}_1$ and $\mathcal{F}_2$ are geometrically isomorphic. In [11], we use this fact, as well as bounds on the number of quasi-orthogonal unit vectors in a finite-dimensional Hilbert space to bound from above the number of geometrically irreducible ℓ-adic sheaves with bounded complexity.

In order to illustrate how this theorem is used, we will state precisely a version of the main theorem of [7] and explain which sums of trace functions arise in the proof.

A holomorphic cusp form of integral weight[2] $k \geq 2$ and level $N \geq 1$ is a holomorphic function

$$f : \mathbf{H} \longrightarrow \mathbf{C}$$

such that

$$f\Big(\frac{az+b}{cz+d}\Big) = (cz+d)^k f(z)$$

for all elements in the subgroup $\Gamma_0(N)$ of elements $\gamma = \begin{pmatrix} a & b \\ c & d \end{pmatrix} \in \mathrm{SL}_2(\mathbf{Z})$ such that N divides c, and furthermore

$$\int_{F_N} |f(z)|^2 y^k \frac{dxdy}{y^2} < +\infty, \tag{4.3}$$

where

$$F_N = \bigcup_{\gamma \in \Gamma_0(N) \backslash \mathrm{SL}_2(\mathbf{Z})} \gamma \cdot F$$

in terms of the fundamental domain F as in Section 1. It follows that $f(z+1) = f(z)$ and therefore f has a Fourier expansion, which holomorphy and the growth condition force to be

$$f(z) = \sum_{n \geq 1} n^{(k-1)/2} \rho_f(n) e(nz)$$

for some coefficients $\rho_f(n) \in \mathbf{C}$ (the normalizing factor $n^{(k-1)/2}$ has the effect of ensuring that $\rho_f(n)$ is bounded in mean-square average).

Theorem 4.3 ([7]). *Let f be a fixed cusp form as above. For any prime p and for any function $K : \mathbf{Z} \to \mathbf{C}$ such that $K(n) = t_{\mathcal{F}}(n \pmod p)$ for some ℓ-adic representation $\mathcal{F}$ modulo p of weight* 0, *we have*

$$\sum_{n \leq X} \rho_f(n) K(n) \ll \mathbf{c}(\mathcal{F})^9 X \Big(1 + \frac{p}{X}\Big)^{1/2} p^{-1/16+\varepsilon}$$

for $X \geq 1$ and any $\varepsilon > 0$, where the implied constant depends only on f and on $\varepsilon > 0$.

The trivial bound for the sum is

$$\Big|\sum_{n \leq X} \rho_f(n) K(n)\Big| \leq \Big(\sum_{n \leq X} |\rho_f(n)|^2\Big)^{1/2} \Big(\sum_{n \leq X} |K(n)|^2\Big)^{1/2} \ll \mathbf{c}(\mathcal{F}) X$$

[2] This notion of weight is not directly related in general to that of weight 0 sheaves.

by the well-known Rankin-Selberg estimate

$$\sum_{n\leq X} |\rho_f(n)|^2 \sim c_f X$$

for some $c_f > 0$ as $X \to \infty$. Thus, assuming that the conductor is bounded by a fixed constant B, our theorem is non-trivial provided X is of size between p (or a bit smaller) and p^A for some fixed A. For the critical case $X = p$, we get a saving of size $p^{-1/16+\varepsilon}$ over the trivial bound.

In particular, if we have an integer $B \geq 1$, and for each prime p a trace function K_p in such a way that the associated representations satisfy $\mathbf{c}(\mathcal{F}_p) \leq B$, then it follows that for p large, there is no correlation between the phases of $\rho_f(n)$ and those of $K_p(n)$.

Slightly more general versions of this theorem apply to the sums (1.3) which arose in Section 1, and combined with the decomposition (3.2) of the functions (1.4), this leads to the proof of the limit relations (1.5) discussed in the motivating problem.

A striking feature of Theorem 4.3 is the universality of the exponent $1/16$ (which can be improved to $1/8$ for "smooth" sums). This is a direct effect of Theorem 4.1 and the universality of the exponent $1/2$ in the right-hand side of (4.1).

We outline the basic strategy of the proof, to indicate where the Riemann Hypothesis comes into play. First, by elementary decompositions, we may assume that the ℓ-adic sheaf $\mathcal{F}$ with trace function K is geometrically irreducible, and by dealing directly with Artin-Schreier sheaves $\mathcal{L}_{\psi(aX)}$, we may assume that $\mathcal{F}$ is a Fourier sheaf. Applying deep results from the theory of automorphic forms (especially the Kuznetsov formula, Hecke theory, and the amplification method) one reduces estimates for the sums in Theorem 4.3 to the study of certain sums of the type

$$\sum_{\boldsymbol{\alpha}\in X} c(\boldsymbol{\alpha})\mathcal{C}(K;\gamma(\boldsymbol{\alpha})) \tag{4.4}$$

where X is a certain finite set of parameters, $c(\boldsymbol{\alpha})$ are complex numbers, $\gamma(\boldsymbol{\alpha})$ is an element of the finite group $\mathrm{PGL}_2(\mathbf{F}_p)$, and the *correlation sums* $\mathcal{C}(K;\gamma(\boldsymbol{\alpha}))$ are defined by

$$\mathcal{C}(\varphi;\gamma) = \sum_{\substack{x\in\mathbf{F}_p\\ \gamma\cdot x\neq\infty}} \overline{\mathrm{FT}(\varphi)(x)}\,\mathrm{FT}(\varphi)(\gamma\cdot x)$$

for any function $\varphi : \mathbf{F}_p \longrightarrow \mathbf{C}$ and $\gamma \in \mathrm{PGL}_2(\mathbf{F}_p)$, which we view as acting on $\mathbf{P}^1(\mathbf{F}_p) = \mathbf{F}_p \cup \{\infty\}$ by the usual action (the same formula as in (1.1)).

From the theory of the Fourier transform, as explained in Section 3.4, we know that $\mathrm{FT}(K)$ is a trace function associated to a geometrically irreducible Fourier sheaf of weight 0 with conductor $\leq 10\mathbf{c}(\mathcal{F})^2$. Furthermore, for any ℓ-adic sheaf modulo p and $\gamma \in \mathrm{PGL}_2(\mathbf{F}_p)$, we have an elementary definition of an ℓ-adic sheaf $\gamma^*\mathcal{F}$ with trace function given by $x \mapsto t_{\mathcal{F}}(\gamma \cdot x)$, and with the same conductor as $\mathcal{F}$. Thus the factor $\mathrm{FT}(K)(\gamma \cdot x)$ is also the trace function of an ℓ-adic sheaf (geometrically irreducible, weight 0) with conductor $\leq 10\mathbf{c}(\mathcal{F})^2$.

It turns out that the reduction procedure implies that good estimates for the original sum follow if we can use in (4.4) a square-root cancellation estimate

$$|\mathcal{C}(K;\gamma(\boldsymbol{\alpha}))| \leq Cp^{1/2} \tag{4.5}$$

for *all* $\boldsymbol{\alpha} \in X$. On the other hand, Theorem 4.1 easily implies that if C is large enough in terms of the conductor of $\mathcal{F}$, we have an inclusion

$$\begin{aligned} &\{\gamma \in \mathrm{PGL}_2(\mathbf{F}_p) \mid |\mathcal{C}(K;\gamma)| > Cp^{1/2}\} \\ \subset &\{\gamma \in \mathrm{PGL}_2(\mathbf{F}_p) \mid \gamma^*\,\mathrm{FT}_\psi(\mathcal{F}) \simeq \mathrm{FT}_\psi(\mathcal{F})\} \end{aligned}$$

(where $\simeq$ denotes geometric isomorphism; we use here the reduction to geometrically irreducible $\mathcal{F}$).

The crucial point is that the right-hand side is a group, which we denote $\mathbf{G}_{\mathrm{FT}_\psi(\mathcal{F})} \subset \mathrm{PGL}_2(\mathbf{F}_p)$ and call the "Möbius group of $\mathrm{FT}_\psi(\mathcal{F})$". This group may well be non-trivial, or relatively large, so that (4.5) cannot in general be expected to hold in all cases.

Using the classification of subgroups of $\mathrm{PGL}_2(\mathbf{F}_p)$, we can nevertheless conclude using the following proposition:

Proposition 4.4. *Let p be a prime number and let $\mathcal{G}$ be a geometrically irreducible ℓ-adic Fourier sheaf of weight* 0 *modulo p. Then, if p is large enough compared to the conductor of $\mathcal{G}$, one of the following properties holds:*

(1) *The Möbius group $\mathbf{G}_{\mathcal{G}}$ contains an element of order p; in this case $\mathcal{G} \simeq \gamma^*\mathcal{L}_{\psi(aX)}$ for some $a \in \bar{\mathbf{F}}_p$ and some element $\gamma \in \mathrm{PGL}_2(\mathbf{F}_p)$, and $\mathbf{G}_{\mathcal{F}}$ is then conjugate in* $\mathrm{PGL}_2(\mathbf{F}_p)$ *to the subgroup*

$$U = \left\{\begin{pmatrix} 1 & t \\ 0 & 1 \end{pmatrix} \mid t \in \mathbf{F}_p\right\}.$$

(2) *The Möbius group $\mathbf{G}_{\mathcal{G}}$ has order coprime to p, and in this case it is contained in the union of at most* 60 *subgroups, each of which is*

either a conjugate of the normalizer in $\mathrm{PGL}_2(\mathbf{F}_p)$ *of the diagonal torus*

$$T = \left\{ \begin{pmatrix} x & 0 \\ 0 & y \end{pmatrix} \mid x, y \in \mathbf{F}_p^\times \right\}$$

or a conjugate of the normalizer in $\mathrm{PGL}_2(\mathbf{F}_p)$ *of a non-split torus*

$$T_1 = \left\{ \begin{pmatrix} a & b \\ \varepsilon b & a \end{pmatrix} \mid a^2 - \varepsilon b^2 \neq 0 \right\},$$

where $\varepsilon \in \mathbf{F}_p^\times$ *is a non-square.*

We can then exploit the fact that the elements of the form $\gamma(\boldsymbol{\alpha})$ are explicit, and from their origin in the analytic steps, they have no particular algebraic structure. In particular, they are seen to *not* be conjugate to elements of the subgroup U in this proposition (for p large enough compared with the conductor), so that in the first case of the proposition (applied to $\mathcal{G} = \mathrm{FT}_\psi(\mathcal{F})$), we have the estimates (4.5) for all $\gamma(\boldsymbol{\alpha})$. If the second case of the proposition applies, on the other hand, we exploit a repulsion argument for each of the finitely many possible conjugates N of the normalizer of a torus to show, roughly, that if one $\gamma(\boldsymbol{\alpha})$ is in N (which may happen) then there can only be extremely few other $\boldsymbol{\alpha}'$ with $\gamma(\boldsymbol{\alpha}') \in \mathbf{G}_{\mathrm{FT}_\psi(\mathcal{F})}$. Such a small set of exceptions to the estimate (4.5) can then be handled.

Example 4.5.

(1) Let

$$K(n) = e\Big(\frac{\bar{n}}{p}\Big),$$

the trace function of the Artin-Schreier sheaf $\mathcal{L}_{\psi(X^{-1})}$. Then $-\mathrm{FT}(K)$ is the trace function of the Kloosterman sheaf $\mathcal{K}\ell_2$ modulo p, and hence

$$\mathcal{C}(K;\gamma) = \sum_{\substack{x\in\mathbf{F}_p \\ cx+d\neq 0}} \mathrm{Kl}_2(x;p)\,\mathrm{Kl}_2\Big(\frac{ax+b}{cx+d};p\Big)$$

for $\gamma = \begin{pmatrix} a & b \\ c & d \end{pmatrix}$ (since Kloosterman sums $\mathrm{Kl}_2(x;p)$ are real numbers). One can show that $\mathbf{G}_{\mathcal{K}\ell_2} = 1$ is the trivial group, and hence there exists a constant $C \geq 1$ such that

$$\Big| \sum_{\substack{x\in\mathbf{F}_p \\ cx+d\neq 0}} \mathrm{Kl}_2(x;p)\,\mathrm{Kl}_2\Big(\frac{ax+b}{cx+d};p\Big) \Big| \leq Cp^{1/2}$$

for all p prime and $\gamma \neq 1$. We will see in the next section some other applications of special cases of this estimate.

(2) For p prime and $n \in \mathbf{F}_p$, define

$$K(n) = \mathrm{Kl}_2(n^2; p) - 1.$$

This is the trace function of an ℓ-adic sheaf modulo p, the symmetric square $\mathcal{F}$ of the pull-back of the Kloosterman sheaf $\mathcal{K}\ell_2$ under the map $x \mapsto x^2$, and this description also shows that the conductor of $\mathcal{F}$ is bounded independently of p. It is a non-trivial fact that $\mathbf{G}_{\mathrm{FT}_\psi(\mathcal{F})}$, in that case, is the subgroup of $\mathrm{PGL}_2(\mathbf{F}_p)$ stabilizing the subset $\{\infty, 0, 4, -4\}$ of $\mathbf{P}^1(\mathbf{F}_p)$, which is a dihedral group of order 8. (More precisely, the inclusion of $\mathbf{G}_{\mathrm{FT}_\psi(\mathcal{F})}$ in this group is elementary, because $\mathcal{F}$ is ramified exactly at these points, and the converse can be checked in different ways, none of which is elementary – maybe the most elegant is to use a result of Deligne and Flicker [5, Cor. 7.7].)

Prior to [7], the only instances of Theorem 4.3 that were considered in the literature (to our knowledge) where $K(n) = e(an/p)$, an additive character, or $K(n) = \chi(n)$, where χ is a non-trivial Dirichlet character modulo p. In the first case, there is an even stronger bound

$$\sum_{n \leq X} \rho_f(n) e\Big(\frac{an}{p}\Big) \ll X^{1/2}(\log X)$$

due to Wilton, valid uniformly for all a modulo p (and in fact we use this estimate directly for $K(n) = e(an/p)$, or equivalently for Artin-Schreier sheaves $\mathcal{L}_{\psi(aX)}$.) The case of a multiplicative character is related to the subconvexity problem for the twisted special values of L-functions $L(f \otimes \chi, 1/2)$ (see for instance [23, Lecture 4] for a survey), and non-trivial estimates were first found by Duke, Friedlander and Iwaniec [6]. The bound in Theorem 4.3 recovers the best known result in terms of the modulus, due to Blomer and Harcos [1] (although the latter deals more generally with characters to all moduli $q \geq 1$, and not only primes). In this case, one sees that $\mathrm{FT}_\psi(\mathcal{L}_\chi) \simeq \mathcal{L}_{\bar\chi}$ and that $\mathbf{G}_{\mathcal{L}_{\bar\chi}}$ is either the diagonal torus T of Proposition 4.4 (2), or its normalizer in $\mathrm{PGL}_2(\mathbf{F}_p)$ (this last case occurring if and only if χ is a real character).

5. Distribution of arithmetic functions in arithmetic progressions

The result of Theorem 4.3 is not only interesting as a statement concerning modular forms. Generalizing the result to encompass Eisenstein series and not only cusp forms, and applying further methods from the analytic study of prime numbers, as well as more general properties of trace functions, we obtained in [8] a striking application to sums over primes (or against the Möbius function).

Theorem 5.1 ([8]). *Let p be a prime number and let $K = t_{\mathcal{F}}$ be the trace function of an ℓ-adic middle-extension $\mathcal{F}$ of weight 0 modulo p such that no geometrically irreducible component of $\mathcal{F}$ is geometrically isomorphic to a tensor product*

$$\mathcal{L}_\psi \otimes \mathcal{L}_\chi$$

where ψ is a possibly trivial additive character and χ a possibly trivial multiplicative character. There exists an absolute constant $B \geq 0$ such that

$$\sum_{n \leq X} \Lambda(n) K(n) \ll \mathbf{c}(\mathcal{F})^B X \Big(1 + \frac{p}{X}\Big)^{1/12} p^{-1/48+\varepsilon},$$

$$\sum_{n \leq X} \mu(n) K(n) \ll \mathbf{c}(\mathcal{F})^B X \Big(1 + \frac{p}{X}\Big)^{1/12} p^{-1/48+\varepsilon}$$

for any $\varepsilon > 0$, where the implied constant depends only on $\varepsilon > 0$. Here Λ denotes the von Mangoldt function and μ the Möbius function.

We remark that the restriction on $\mathcal{F}$ is, with current techniques, necessary: an estimate of this quality for $K(n) = \chi(n)$ would imply a non-trivial zero-free strip in the critical strip for the Dirichlet L-function $L(s, \chi)$. This assumption holds however in many cases, for instance whenever $\mathcal{F}$ is geometrically irreducible with rank at least 2, or if $\mathcal{F}$ is ramified at some point $x \in \mathbf{P}^1(\bar{\mathbf{F}}_p) - \{0, \infty\}$, or if $\mathcal{F} = \mathcal{L}_{\chi(f)}$ with χ non-trivial and $f \in \mathbf{F}_p(X)$ not a monomial, or if $\mathcal{F} = \mathcal{L}_{\psi(f)}$ with ψ non-trivial and $f \in \mathbf{F}_p(X)$ not a polynomial of degree ≤ 1.

The interest of this theorem is when X is close to p (for X much larger, one can use periodicity instead). Prior to [8], only the following cases had been studied in this range:

(1) When $K(n) = \chi(f(n))$, where f is a polynomial of degree ≤ 2 which is not a monomial (Vinogradov, Karatsuba);

(2) When $K(n) = e(f(n)/p)$ for certain rational functions $f \in \mathbf{Q}(X)$ (Vinogradov, Fouvry–Michel [14], ...)

The new ingredient concerning trace functions in the proof of this theorem is the following general estimate:

Theorem 5.2 ([8]). *Let p be a prime and let $K = t_{\mathcal{F}}$ be the trace function of an ℓ-adic middle-extension sheaf $\mathcal{F}$ of weight 0 modulo p such* *that no geometrically irreducible component of $\mathcal{F}$ is geometrically isomorphic to a tensor product*

$$\mathcal{L}_{\psi} \otimes \mathcal{L}_{\chi}$$

where ψ is a possibly trivial additive character and χ a possibly trivial multiplicative character. Let $\alpha = (\alpha(m))$ and $\beta = (\beta(n))$ be sequences of complex numbers supported on $M/2 \leq m \leq M$ and $N/2 \leq n \leq N$ respectively for some $M, N \geq 1$. There exists an absolute constant $B \geq 0$ such that we have

$$\sum_{m}\sum_{n} \alpha(m)\beta(n)K(mn)$$
$$\ll \mathbf{c}(\mathcal{F})^{B} \|\alpha\| \|\beta\| (MN)^{1/2} \Big(\frac{1}{p^{1/4}} + \frac{1}{M^{1/2}} + \frac{p^{1/4}(\log p)^{1/2}}{N^{1/2}} \Big),$$

where the implied constant is absolute.

The basic idea of the proof is classical in analytic number theory: one reduces quickly using the Cauchy-Schwarz inequality to proving that

$$\Big| \sum_{x \in \mathbf{F}_p} K(x)\overline{K(ax)} e\Big(\frac{bx}{p}\Big) \Big| \ll \mathbf{c}(\mathcal{F})^{B} p^{1/2}$$

for all $(a, b) \in \mathbf{F}_p^{\times} \times \mathbf{F}_p$, with *at most* $\mathbf{c}(\mathcal{F})^{B}$ exceptions, where B and the implied constant are absolute.

But the Plancherel formula gives

$$\begin{aligned} \sum_{x \in \mathbf{F}_p} K(x)\overline{K(ax)} e\Big(\frac{bx}{p}\Big) &= \sum_{t \in \mathbf{F}_p} \mathrm{FT}(K)(t)\overline{\mathrm{FT}(K)(-at+b)} \\ &= \mathcal{C}\Big(K; \begin{pmatrix} -a & b \\ 0 & 1 \end{pmatrix}\Big) \end{aligned} \tag{5.1}$$

so that the sums in this theorem are special cases of the correlation sums $\mathcal{C}(K;\gamma)$ of the previous section, for γ restricted to the subgroup B_p of upper-triangular matrices in $\mathrm{PGL}_2(\mathbf{F}_p)$ (interestingly, one of the properties of the decomposition (4.4) used in the proof of Theorem 4.3 is that $\gamma(\boldsymbol{\alpha})$ is *not* upper-triangular in that case!) We then only need to check that, under the assumptions of Theorem 5.2, the intersection of the group $\mathbf{G}_{\mathrm{FT}_{\psi}(\mathcal{F})}$ with B_p has size bounded by $\mathbf{c}(\mathcal{F})^{B}$ for some absolute constant B. This is done using some analysis of the group $\mathbf{G}_{\mathcal{F}}$ (in particular, the non-obvious fact that it is the group of $\mathbf{F}_p$-rational points of an algebraic subgroup of $\mathbf{G}_{\mathcal{F}}$.)

The general estimate of Theorem 5.2 has further applications. One which is dear to our heart is found in [9]: combining it with versions of the Voronoi summation formula and with other tools from [8], we improve significantly the exponent of distribution for the ternary divisor function d_3 in arithmetic progression. We recall that

$$d_3(n) = \sum_{abc=n} 1,$$

so that the Dirichlet generating series of d_3 is

$$\sum_{n\geq 1} d_3(n)n^{-s} = \zeta(s)^3$$

for $\mathrm{Re}(s) > 1$.

Theorem 5.3 ([9]). *Let p be a prime number, and let a be an integer coprime to p. Let $\varepsilon > 0$ be a positive number. For all X such that $p \leq X^{1/2+1/46-\varepsilon}$, we have*

$$\Big| \sum_{\substack{n\leq X \\ n\equiv a \ (\mathrm{mod}\ p)}} d_3(n) - \frac{1}{p-1}\sum_{n\leq X} d_3(n)\Big| \ll \frac{X}{p}\frac{1}{(\log X)^A}$$

for any $A \geq 1$, where the implied constant depends on ε and A.

The essential qualitative point is that the exponent $1/2+1/46$ is beyond $1/2$, which is the limit where a result like this would almost trivially follow from the Generalized Riemann Hypothesis for Dirichlet characters. Going beyond $1/2$ in this problem was first achieved by Friedlander and Iwaniec [15], whose result was improved by Heath-Brown [16]. Our own result, although slightly less general (in that we only consider prime moduli p instead of all $q \geq 1$), is another significant improvement, and most importantly in our mind, the proof is rather straightforward in principle when using the results of [8]. In fact, the only specific trace functions modulo p that we use in the proof are given by

$$K(n) = \mathrm{Kl}_3(an; p)$$

for some $a \in \mathbf{F}_p^\times$. In particular, we make use of the following estimate, which was already used (implicitly) by Friedlander-Iwaniec and Heath-Brown:

Theorem 5.4 (Correlation of hyper-Kloosterman sums). *Let p be a prime and $(a, b) \in \mathbf{F}_p^\times \times \mathbf{F}_p$. There exists a constant $C \geq 1$, independent of p and a, such that*

$$\Big|\sum_{x \in \mathbf{F}_p^\times} \mathrm{Kl}_3(x; p)\overline{\mathrm{Kl}_3(ax; p)}e\Big(\frac{bx}{p}\Big)\Big| \leq C\sqrt{p}$$

for $(a, b) \neq (1, 0)$.

We see from (5.1) that this can be derived from the existence of (and conductor bound for) Kloosterman sheaves and the fact that the group $\mathbf{G}_{\mathrm{FT}_\psi(\mathcal{K}\ell_3)} \subset \mathrm{PGL}_2(\mathbf{F}_p)$ is trivial.

Previously, this exponential sum was handled by Friedlander and Iwaniec (and by Heath-Brown) by writing

$$\sum_{x \in \mathbf{F}_p^\times} \mathrm{Kl}_3(x; p)\overline{\mathrm{Kl}_3(ax; p)}e\Big(\frac{bx}{p}\Big) = \sum_{t \neq 0, -b} \mathrm{Kl}_2\Big(\frac{1}{t}; p\Big)\,\mathrm{Kl}_2\Big(\frac{a}{t+b}; p\Big) - \frac{1}{p^2}$$

(by a simple computation), and using a bound for the sum on the right-hand side, which was itself proved by Bombieri and Birch in the Appendix to [15].

One may observe that the sum on the right-hand side also arises as a correlation sum. Indeed, for $K(n) = e(\bar{n}/p)$ as in Example 4.5 (1), if we take

$$\gamma = \begin{pmatrix} a & 0 \\ b & 1 \end{pmatrix}$$

for $(a, b) \in \mathbf{F}_p^\times \times \mathbf{F}_p$, then a simple change of variable in the definition leads to the identity

$$\mathcal{C}(K; \gamma) = \sum_{\substack{t \in \mathbf{F}_p \\ x \neq -b}} \mathrm{Kl}_2\Big(\frac{1}{t}; p\Big)\,\mathrm{Kl}_2\Big(\frac{a}{t+b}; p\Big).$$

Thus Theorem 5.4 follows also from the fact that the group $\mathbf{G}_{\mathrm{FT}_\psi(\mathcal{K}\ell_2)}$ is trivial, which is not very difficult to prove. Interestingly, other correlations sums attached to the same sheaf appeared in other papers in the literature: we are aware of its occurrence in works of Pitt [25] and Munshi [24].

References

[1] V. Blomer and G. Harcos, *Hybrid bounds for twisted L-functions*, J. reine und angew. Mathematik **621** (2008), 53–79.

[2] N. BOURBAKI, "Fonctions d'une Variable Réelle", Paris, Hermann, 1976.

[3] P. DELIGNE, "Cohomologie Étale", S.G.A $4\frac{1}{2}$, L.N.M., Vol. 569, Springer Verlag, 1977.

[4] P. DELIGNE, *La conjecture de Weil, II*, Publ. Math. IHÉS **52** (1980), 137–252.

[5] P. DELIGNE and Y. Z. FLICKER, *Counting local systems with principal unipotent local monodromy*, Ann. of Math. (2) **178** (2013), no 3, 921–982.

[6] W. D. DUKE, J. FRIEDLANDER and H. IWANIEC, *Bounds for automorphic L-functions*, Invent. Math. **112** (1993), 1–8.

[7] É. FOUVRY, E. KOWALSKI and PH. MICHEL, *Algebraic twists of modular forms and Hecke orbits*, preprint (2012), arXiv:1207.0617.

[8] É. FOUVRY, E. KOWALSKI and PH. MICHEL, *Algebraic trace weights over the primes*, Duke Math. J. **163** (2014), 1683–1736.

[9] É. FOUVRY, E. KOWALSKI and PH. MICHEL, *On the exponent of distribution of the ternary divisor function*, Mathematika, available online at doi:10.1112/S0025579314000096

[10] É. FOUVRY, E. KOWALSKI and PH. MICHEL, *An inverse theorem for Gowers norms of trace functions over* $\mathbf{F}_p$, Math. Proc. Cambridge Phil. Soc. **155** (2013), 277–295.

[11] É. FOUVRY, E. KOWALSKI and PH. MICHEL, *Counting sheaves with spherical codes*, Math. Res. Letters **20** (2013), 305–323.

[12] É. FOUVRY, E. KOWALSKI and PH. MICHEL, *The sliding sum method for short exponential sums*, preprint (2013), arXiv:1307.0135.

[13] É. FOUVRY, E. KOWALSKI and PH. MICHEL, *On the conductor of cohomological transforms*, preprint (2013), arXiv:1310.3603.

[14] É. FOUVRY and PH. MICHEL, *Sur certaines sommes d'exponentielles sur les nombres premiers*, Ann. Sci. École Norm. Sup. (4) **31** (1998), 93–130.

[15] J. B. FRIEDLANDER and H. IWANIEC, *Incomplete Kloosterman sums and a divisor problem*, (With an appendix by Bryan J. Birch and Enrico Bombieri), Ann. of Math. (2) **121** (1985), 319–350.

[16] D. R. HEATH-BROWN, *The divisor function $d_3(n)$ in arithmetic progressions*, Acta Arith. **47** (1986), no. 1, 29–56.

[17] J. HEINLOTH, B-C. NGÔ and Z. YUN, *Kloosterman sheaves for reductive groups*, Ann. of Math. (2) **177** (2013), no. 1, 241–310.

[18] H. IWANIEC and E. KOWALSKI, "Analytic Number Theory", A.M.S. Coll. Publ., Vol. 53, 2004.

[19] N. M. KATZ, "Gauss Sums, Kloosterman Sums and Monodromy Groups", Annals of Math. Studies 116, Princeton Univ. Press, 1988.

[20] N. M. KATZ, "Exponential Sums and Differential Equations", Annals of Math. Studies, Vol. 124, Princeton Univ. Press, 1990.

[21] E. KOWALSKI, *Poincaré and analytic number theory*, In: "The scientific legacy of Poincaré", É. Charpentier, É. Ghys and A. Lesne (eds.), Hist. Math., Vol. 36, Amer. Math. Soc., 2010.

[22] G. LAUMON, *Transformation de Fourier, constantes d'équations fonctionnelles et conjecture de Weil*, Publ. Math. IHÉS **65** (1987), 131–210.

[23] PH. MICHEL, *Analytic number theory and families of automorphic L-functions*, In: "Automorphic Forms and Applications", IAS/Park City Math. Ser., Vol. 12, Amer. Math. Soc., 2007, 181–295.

[24] R. MUNSHI, *Shifted convolution sums for* $GL(3) \times GL(2)$, preprint (2012), arXiv:1202.1157.

[25] N. PITT, *On shifted convolutions of* $\zeta(s)^3$ *with automorphic L-functions*, Duke Math. J. **77** (1995), no. 2, 383–406.

Topological methods in algebraic geometry

Fabrizio Catanese

Contents

The present work took place in the realm of the DFG Forschergruppe 790 "Classification of algebraic surfaces and compact complex manifolds" and of the ERC advanced grant 340258 TADMICAMT.

Prologue

Let me begin by citing Hermann Weyl ([93, p. 500]):

'In these days the angel of topology and the devil of abstract algebra fight for the soul of each individual mathematical domain'.

My motivation for this citation is first of all a practical reflexion on the primary role played by the field of Topology in the mathematics of the 20-th century, and the danger that among algebraic geometers this great heritage and its still vivid current interest may be not sufficiently considered.

Second, the word 'soul' used by Weyl reminds us directly of the fact that mathematics is one of the pillars of scientific culture, and that some philosophical discussion about its role in society is deeply needed. [1]

Also, dozens of years dominated by neo liberism, and all the rest, have brought many of us to accept the slogan that mathematics is a key-technology. So, the question which is too often asked is: 'for which immediate purposes is this good for?'[2] Instead of asking: 'how beautiful, important or enriching is this theory?', or 'how do all these theories contribute to deep knowledge and wisdom, and to broad scientific progress?'

While it is of course true that mathematics is extremely useful for the advancement of society and the practical well being of men, yet I would wish that culture and mathematics should be highly respected and supported, without the need of investing incredible amounts of energy devoted to make it survive. Our energy should better be reserved to the major task of making mathematical culture more unified, rather than a Babel tower where adepts of different disciplines can hardly talk to each other.

Thus, in a way, one should conclude trying to underline the fruitful interactions among several fields of mathematics, and thus paraphrase the motto by Weyl by asking: 'How can the angel of topology live happily with the devil of abstract algebra?'.

Now, the interaction of algebraic geometry and topology has been such, in the last three centuries, that it is often difficult to say when does a result belong to one discipline or to the other, the archetypical example being the Bézout theorem, first conceived through geometrical ideas, and later clarified through topology and through algebra.

---

[1] Mathematics privileges problem solving and critical thinking versus passive acceptance of dogmatic 'truths'. And the peaceful survival of our current world requires men to lose their primitive nature and mentality and to become culturally more highly developed.

[2] And too often this is only measured by monetary or immediate financial success.

Thus, the ties are so many that I will have to soon converge towards my personal interests. I shall mostly consider moduli theory as the fine part of classification theory of complex varieties: and I shall try to show how in some lucky cases topology helps also for the fine classification, allowing the study of the structure of moduli spaces: as we have done quite concretely in several papers ([10–13,15,16]). Finally I shall present how the theory of moduli, guided by topological considerations, gives in return important information on the Galois group of the field $\bar{\mathbb{Q}}$.

For a broader treatment, I refer the reader to the article [38], to which this note is an invitation.

1. Applications of algebraic topology: non existence and existence of continuous maps

Algebraic topology flourished from some of its applications, inferring the non existence of certain continuous maps from the observation that their existence would imply the existence of homomorphisms satisfying algebraic properties which are manifestly impossible to be verified. The most famous such examples are Brouwer's fixed point theorem, and the theorem of Borsuk-Ulam.

Theorem 1.1 (Brouwer's fixed point theorem). *Every continuous self map $f : D^n \to D^n$, where $D^n = \{x \in \mathbb{R}^n | |x| \leq 1\}$ is the unit disk, has a fixed point,* i.e., *there is a $x \in D$ such that $f(x) = x$.*

The proof is by contradiction:

1. Assuming that $f(x) \neq x \; \forall x$, let $\phi(x)$ be the intersection of the boundary S^{n-1} of D^n with the half line stemming from $f(x)$ in the direction of x; ϕ would be a continuous map
$$\phi : D^n \to S^{n-1}, \; s.t. \; \phi|_{S^{n-1}} = \mathrm{Id}_{S^{n-1}} .$$
2. *i.e.*, we would have a sequence of two continuous maps (ι is the inclusion) whose composition $\phi \circ \iota$ is the identity
$$\iota : S^{n-1} \to D^n, \; \phi : D^n \to S^{n-1}.$$
3. One uses then the **covariant functoriality** of reduced homology groups $H_i(X, \mathbb{Z})$: to each continuous map $f : X \to Y$ of topological spaces is associated a homomorphism of abelian groups $H_i(f, \mathbb{Z}) : H_i(X, \mathbb{Z}) \to H_i(Y, \mathbb{Z})$, and in such a way that to a composition $f \circ g$ is associated the composition of the corresponding homomorphisms. That is,
$$H_i(f \circ g, \mathbb{Z}) = H_i(f, \mathbb{Z}) \circ H_i(g, \mathbb{Z}).$$
Moreover, to the identity is associated the identity.

4. The key point is (one observes that the disc is contractible) to show that the reduced homology groups

$$H_{n-1}(S^{n-1}, \mathbb{Z}) \cong \mathbb{Z}, \quad H_{n-1}(D^n, \mathbb{Z}) = 0.$$

The functoriality of the homology groups, since $\phi \circ \iota = \mathrm{Id}_{S^{n-1}}$, would imply $0 = H_{n-1}(\phi, \mathbb{Z}) \circ H_{n-1}(\iota, \mathbb{Z}) = H_{n-1}(\mathrm{Id}_{S^{n-1}}, \mathbb{Z}) = \mathrm{Id}_{\mathbb{Z}}$, the desired contradiction.

The cohomology algebra is used instead for the Borsuk-Ulam theorem.

Theorem 1.2 (Borsuk-Ulam theorem). *There exists no odd continuous function $F : S^n \to S^m$ for $n > m$ (F is odd means that $F(-x) = -F(x), \forall x$).*

Here there are two ingredients, the main one being the cohomology algebra, and its contravariant functoriality: to any continuous map $f : X \to Y$ there corresponds an algebra homomorphism

$$f^* : H^*(Y, R) = \oplus_{i=0}^{\dim(Y)} H^i(Y, R) \to H^*(X, R),$$

for any ring R of coefficients.

In our case one takes as $X := \mathbb{P}^n_{\mathbb{R}} = S^n/\{\pm 1\}$, similarly $Y := \mathbb{P}^m_{\mathbb{R}} = S^m/\{\pm 1\}$ and lets f be the continuous map induced by F.

One needs to show that, choosing $R = \mathbb{Z}/2\mathbb{Z}$, then the cohomology algebra of real projective space is a truncated polynomial algebra, namely:

$$H^*(\mathbb{P}^n_{\mathbb{R}}, \mathbb{Z}/2\mathbb{Z}) \cong (\mathbb{Z}/2\mathbb{Z})[\xi_n]/(\xi_n^{n+1}).$$

The other ingredient consists in showing that

$$f^*([\xi_m]) = [\xi_n],$$

$[\xi_m]$ denoting the residue class in the quotient algebra.

One gets then the desired contradiction since, if $n > m$,

$$0 = f^*(0) = f^*([\xi_m]^{m+1}) = f^*([\xi_m])^{m+1} = [\xi_n]^{m+1} \neq 0.$$

Notice that up to now we have mainly used that f is a continuous map $f := \mathbb{P}^n_{\mathbb{R}} \to \mathbb{P}^m_{\mathbb{R}}$, while precisely in order to obtain that $f^*([\xi_m]) = [\xi_n]$ we must make use of the hypothesis that f is induced by an odd function F.

This property can be interpreted as the property that F yields a commutative diagram

$$\begin{array}{ccc} S^n & \to & S^m \\ \downarrow & & \downarrow \\ \mathbb{P}^n_{\mathbb{R}} & \to & \mathbb{P}^m_{\mathbb{R}} \end{array}$$

which exhibits the two sheeted covering of $\mathbb{P}^n_{\mathbb{R}}$ by S^n as the pull-back of the analogous two sheeted cover for $\mathbb{P}^m_{\mathbb{R}}$. Now, as we shall digress soon, any such two sheeted covering is given by a homomorphism of $H_1(X, \mathbb{Z}/2\mathbb{Z}) \to \mathbb{Z}/2\mathbb{Z}$, *i.e.*, by an element in $H^1(X, \mathbb{Z}/2\mathbb{Z})$, and this element is trivial if and only if the covering is trivial (that is, homeomorphic to $X \times (\mathbb{Z}/2\mathbb{Z})$, in other words a disconnected cover).

This shows that the pull back of the cover, which is nontrivial, corresponds to $f^*([\xi_m])$ and is nontrivial, hence $f^*([\xi_m]) = [\xi_n]$.

In this way the proof is accomplished.

Algebraic topology attaches to a good topological space homology groups $H_i(X, R)$, which are covariantly functorial, a cohomology algebra $H^*(X, R)$ which is contravariantly functorial, and these groups can be calculated, by virtue of the Mayer-Vietoris exact sequence and of excision (see any textbook), by chopping the space in smaller pieces. In particular, these groups vanish when $i > \dim(X)$.

To X are also attached the homotopy groups $\pi_i(X)$.

Definition 1.3.

(1) Let $f, g : X \to Y$ be continuous maps. Then f and g are said to be homotopic (one writes $f \sim g$) if there is continuous map $F : X \times [0, 1] \to Y$ such that $f(x) = F(x, 0)$ and $g(x) = F(x, 1)$. Similar definition for maps of pairs $f, g : (X, X') \to (Y, Y')$, which means that $X' \subset X$ is mapped to $f(X') \subset Y' \subset Y$.

(2) $[X, Y]$ is the set of homotopy classes of continuous maps $f : X \to Y$.

(3) $\pi_i(X, x_0) := [(S^i, e_1), (X, x_0)]$ is a group for $i \geq 1$, abelian for $i \geq 2$, and independent of the point $x_0 \in X$ if X is path-connected.

(4) X is said to be homotopy equivalent to Y ($X \sim Y$) if and only if there are continuous maps $f : X \to Y$, $g : Y \to X$ such that $f \circ g$ and $g \circ f$ are both homotopic to the identity (of Y, resp. of X).

The common feature is that homotopic maps induce the same homomorphisms on homology, cohomology, and homotopy.

We are, for our purposes, more interested in the more mysterious homotopy groups, which, while not necessarily vanishing for $i > \dim(X)$, enjoy however a fundamental property.

Recall the definition due to Whitney and Steenrod ([88]) of a fibre bundle. In the words of Steenrod, the notion of a fibre bundle is a weakening of the notion of a product, since a product $X \times Y$ has two continuous projections $p_X : X \times Y \to X$, and $p_Y : X \times Y \to Y$, while a fibre bundle E over B with fibre F has only one projection, $p = p_B : E \to B$ and its similarity to a product lies in the fact that for each point $x \in B$ there is an open set U containing x, and a homeomorphism of $p_B^{-1}(U) \cong U \times F$ compatible with both projections onto U.

The fundamental property of fibre bundles is that there is a long exact sequence of homotopy groups

$$\ldots \to \pi_i(F) \to \pi_i(E) \to \pi_i(B) \to \pi_{i-1}(F) \to \pi_{i-1}(E) \to \pi_{i-1}(B) \to \ldots$$

where one should observe that $\pi_i(X)$ is a group for $i \geq 1$, an abelian group for $i \geq 2$, and for $i = 0$ is just the set of arc-connected components of X (we assume the spaces to be good, that is, locally arcwise connected, semilocally simply connected, see [55], and, most of the times, connected).

The special case where the fibre F has the discrete topology is the case of a **covering space**, which is called the universal covering if moreover $\pi_1(E)$ is trivial.

Special mention deserves the following more special case.

Definition 1.4. Assume that E is arcwise connected, contractible (hence all homotopy groups $\pi_i(E)$ are trivial), and that the fibre F is discrete, so that all the higher homotopy groups $\pi_i(B) = 0$ for $i \geq 2$, while $\pi_1(B) \cong \pi_0(F) = F$.

Then one says that B is a classifying space $K(\pi, 1)$ for the group $\pi = \pi_1(B)$.

In general, given a group π, a CW complex B is said to be a $K(\pi, 1)$ if $\pi_i(B) = 0$ for $i \geq 2$, while $\pi_1(B) \cong \pi$.

Example 1.5. The easiest examples are the following ones:

1. the real torus $T^n := \mathbb{R}^n/\mathbb{Z}^n$ is a classifying space $K(\mathbb{Z}^n, 1)$ for the group $\pi = \mathbb{Z}^n$;
2. a classifying space $K(\mathbb{Z}/2\mathbb{Z}, 1)$ is given by the inductive limit $\mathbb{P}^\infty_\mathbb{R} := \lim_{n\to\infty} \mathbb{P}^n_\mathbb{R}$.

These classifying spaces, although not unique, are unique up to **homotopy-equivalence** (we use the notation $X \sim_{h.e.} Y$ to denote homotopy equivalence, defined above and meaning that there exist continuous maps $f : X \to Y$, $g : Y \to X$ such that both compositions $f \circ g$ and $g \circ f$ are homotopic to the identity).

Therefore, given two classifying spaces for the same group, they not only do have the same homotopy groups, but also the same homology and cohomology groups. Thus the following definition is well posed.

Definition 1.6. Let Γ be a finitely presented group, and let $B\Gamma$ be a classifying space for Γ: then the homology and cohomology groups and algebra of Γ are defined as

$$H_i(\Gamma, \mathbb{Z}) := H_i(B\Gamma, \mathbb{Z}),$$
$$H^i(\Gamma, \mathbb{Z}) := H^i(B\Gamma, \mathbb{Z}),$$
$$H^*(\Gamma, \mathbb{Z}) := H^*(B\Gamma, \mathbb{Z}),$$

and similarly for other rings of coefficients instead of $\mathbb{Z}$.

We now come to the other side: algebraic topology is not only useful to detect the non existence of certain continuous maps, it is also used to assert the existence of certain continuous maps.

Indeed classifying spaces, even if often quite difficult to construct explicitly, are very important because they guarantee the existence of continuous maps! We have more precisely the following (*cf.* [87, Theorem 9, page 427, and Theorem 11, page 428]).

Theorem 1.7. *Let Y be a 'nice' topological space,* i.e., *Y is homotopy-equivalent to a CW-complex, and let X be a nice space which is a $K(\pi, 1)$ space: then, choosing base points $y_0 \in Y, x_0 \in X$, one has a bijective correspondence*

$$[(Y, y_0), (X, x_0)] \cong \mathrm{Hom}(\pi_1(Y, y_0), \pi_1(X, x_0)), [f] \mapsto \pi_1(f),$$

where $[(Y, y_0), (X, x_0)]$ denotes the set of homotopy classes $[f]$ of continuous maps $f : Y \to X$ such that $f(y_0) = x_0$ (and where the homotopies $F(y, t)$ are also required to satisfy $F(y_0, t) = x_0, \ \forall t \in [0, 1]$).

In particular, the free homotopy classes $[Y, X]$ of continuous maps are in bijective correspondence with the conjugacy classes of homomorphisms $\mathrm{Hom}(\pi_1(Y, y_0), \pi)$ *(conjugation is here inner conjugation by* $\mathrm{Inn}(\pi)$ *on the target).*

While topology deals with continuous maps, when dealing with manifolds more regularity is wished for. For instance, when we choose for Y a differentiable manifold M, and the group π is abelian and torsion free, say $\pi = \mathbb{Z}^r$, then a more precise incarnation of the above theorem is given by the De Rham theory.

We have indeed the following proposition.

Proposition 1.8. *Let Y be a differentiable manifold, and let X be a differentiable manifold that is a $K(\pi, 1)$ space: then, choosing base points $y_0 \in Y, x_0 \in X$, one has a bijective correspondence*

$$[(Y, y_0), (X, x_0)]^{\mathrm{diff}} \cong \mathrm{Hom}(\pi_1(Y), \pi), [f] \mapsto \pi_1(f),$$

where $[(Y, y_0), (X, x_0)]^{\mathrm{diff}}$ denotes the set of differential homotopy classes $[f]$ of differentiable maps $f : Y \to X$ such that $f(y_0) = x_0$.

Remark 1.9. In the case where X is a torus $T^r = \mathbb{R}^r/\mathbb{Z}^r$, then f is obtained as the projection onto T^r of

$$\tilde{\phi}(y) := \int_{y_0}^{y} (\eta_1, \ldots, \eta_r), \eta_j \in H^1(Y, \mathbb{Z}) \subset H^1_{DR}(Y, \mathbb{R}).$$

Here η_j is indeed a closed 1-form, representing a certain De Rham cohomology class with integral periods (*i.e.*, $\int_\gamma \eta_j = \varphi(\gamma) \in \mathbb{Z}, \forall \gamma \in \pi_1(Y)$). Therefore f is defined by $\int_{y_0}^{y}(\eta_1, \ldots, \eta_r) \mod (\mathbb{Z}^r)$. Moreover, changing η_j with another form $\eta_j + dF_j$ in the same cohomology class, one finds a homotopic map, since $\int_{y_0}^{y}(\eta_j + tdF_j) = \int_{y_0}^{y}(\eta_j) + t(F_j(y) - F_j(y_0))$.

In algebraic geometry, the De Rham theory leads to the theory of Albanese varieties, which can be understood as dealing with the case where G is free abelian and the classifying maps are holomorphic.

Before we mention other results concerning higher regularity of the classifying maps, we shall now give the basic examples of projective varieties that are classifying spaces.

2. Projective varieties which are $K(\pi, 1)$

The following are the easiest examples of projective varieties which are $K(\pi, 1)$'s.

(1) Projective curves C of genus $g \geq 2$.
By the **Uniformization theorem**, these have the Poincaré upper half plane $\mathcal{H} := \{z \in \mathbb{C} | Im(z) > 0\}$ as universal covering, hence they are compact quotients $C = \mathcal{H}/\Gamma$, where $\Gamma \subset \mathbb{P}SL(2, \mathbb{R})$ is a discrete subgroup isomorphic to the fundamental group of C, $\pi_1(C) \cong \pi_g$. Here

$$\pi_g := \langle \alpha_1, \beta_1, \ldots \alpha_g, \beta_g | \Pi_1^g [\alpha_i, \beta_i] = 1 \rangle$$

contains no elements of finite order.
Moreover, the complex orientation of C induces a standard generator $[C]$ of $H_2(C, \mathbb{Z}) \cong \mathbb{Z}$, the so-called fundamental class.

(2) AV : = Abelian varieties.
More generally, a complex torus $X = \mathbb{C}^g/\Lambda$, where Λ is a discrete subgroup of maximal rank (isomorphic then to $\mathbb{Z}^{2g}$), is a Kähler classifying space $K(\mathbb{Z}^{2g}, 1)$, the Kähler metric being induced by the translation invariant Euclidean metric $\frac{i}{2} \sum_1^g dz_j \otimes d\overline{z_j}$.
For $g = 1$ one gets in this way all projective curves of genus $g = 1$; but, for $g > 1$, X is in general not projective: it is projective, and

called then an Abelian variety, if it satisfies the Riemann bilinear relations. These amount to the existence of a positive definite Hermitian form H on $\mathbb{C}^g$ whose imaginary part A (*i.e.*, $H = S + iA$), takes integer values on $\Lambda \times \Lambda$. In modern terms, there exists a positive line bundle L on X, with Chern class $A \in H^2(X, \mathbb{Z}) = H^2(\Lambda, \mathbb{Z}) = \wedge^2(\mathrm{Hom}(\Lambda, \mathbb{Z}))$, whose curvature form, equal to H, is positive (the existence of a positive line bundle on a compact complex manifold X implies that X is projective algebraic, by Kodaira's theorem, [67]).

(3) LSM : = Locally symmetric manifolds.
These are the quotients of a **bounded symmetric domain** $\mathcal{D}$ by a cocompact discrete subgroup $\Gamma \subset \mathrm{Aut}(\mathcal{D})$ acting freely. Recall that a bounded symmetric domain $\mathcal{D}$ is a bounded domain $\mathcal{D} \subset\subset \mathbb{C}^n$ such that its group $\mathrm{Aut}(\mathcal{D})$ of biholomorphisms contains, for each point $p \in \mathcal{D}$, a holomorphic automorphism σ_p such that $\sigma_p(p) = p$, and such that the derivative of σ_p at p is equal to $-Id$. This property implies that σ is an involution (*i.e.*, it has order 2), and that $\mathrm{Aut}(\mathcal{D})^0$ (the connected component of the identity) is transitive on $\mathcal{D}$, and one can write $\mathcal{D} = G/K$, where G is a connected Lie group, and K is a maximal compact subgroup.
The two important properties are:
(3.1) $\mathcal{D}$ splits uniquely as the product of irreducible bounded symmetric domains.
(3.2) each such $\mathcal{D}$ is contractible.
Bounded symmetric domains were classified by Elie Cartan in [24], and there is only a finite number of them (up to isomorphism) for each dimension n.
Recall the notation for the simplest irreducible domains:

(i) $I_{n,p}$ is the domain $\mathcal{D} = \{Z \in \mathrm{Mat}(n, p, \mathbb{C}) : \mathrm{I}_p -{}^t Z \cdot \overline{Z} > 0\}$.
(ii) II_n is the intersection of the domain $I_{n,n}$ with the subspace of skew symmetric matrices.
(iii) III_n is instead the intersection of the domain $I_{n,n}$ with the subspace of symmetric matrices.

We refer the reader to [60], Theorem 7.1, page 383 and exercise D, pages 526-527, for a list of these irreducible bounded symmetric domains.
In the case of type III domains, the domain is biholomorphic to the Siegel's upper half space:

$$\mathcal{H}_g := \{\tau \in \mathrm{Mat}(g, g, \mathbb{C}) | \tau = {}^t\tau, \mathrm{Im}(\tau) > 0\},$$

a generalisation of the upper half-plane of Poincaré.

(4) A particular, but very explicit case of locally symmetric manifolds is given by the VIP : = Varieties isogenous to a product.
These were studied in [29], and they are defined as quotients

$$X = (C_1 \times C_2 \times \cdots \times C_n)/G$$

of the product of projective curves C_j of respective genera $g_j \geq 2$ by the action of a finite group G acting freely on the product.
In this case the fundamental group of X is not so mysterious and fits into an exact sequence

$$1 \to \pi_1(C_1 \times C_2 \times \cdots \times C_n) \cong \pi_{g_1} \times \cdots \times \pi_{g_n} \to \pi_1(X) \to G \to 1.$$

Such varieties are said to be of the **unmixed type** if the group G does not permute the factors, *i.e.*, there are actions of G on each curve such that

$$\gamma(x_1, \ldots, x_n) = (\gamma x_1, \ldots, \gamma x_n), \forall \gamma \in G.$$

Equivalently, each individual subgroup π_{g_j} is normal in $\pi_1(X)$.

(5) Hyperelliptic surfaces: these are the quotients of a complex torus of dimension 2 by a finite group G acting freely, and in such away that the quotient is not again a complex torus.
These surfaces were classified by Bagnera and de Franchis ([4], see also [51] and [5]) and they are obtained as quotients $(E_1 \times E_2)/G$ where E_1, E_2 are two elliptic curves, and G is an abelian group acting on E_1 by translations, and on E_2 effectively and in such a way that $E_2/G \cong \mathbb{P}^1$.

(6) In higher dimension we define the Generalized Hyperelliptic Varieties (GHV) as quotients A/G of an Abelian Variety A by a finite group G acting freely, and with the property that G is not a subgroup of the group of translations. Without loss of generality one can then assume that G contains no translations, since the subgroup G_T of translations in G would be a normal subgroup, and if we denote $G' = G/G_T$, then $A/G = A'/G'$, where A' is the Abelian variety $A' := A/G_T$.
We proposed instead the name **Bagnera-de Franchis (BdF) Varieties** for those quotients $X = A/G$ were G contains no translations, and G is a cyclic group of order m, with generator g (observe that, when A has dimension $n = 2$, the two notions coincide, thanks to the classification result of Bagnera-De Franchis in [4]).
A concrete description of such Bagnera-De Franchis varieties is given in [38].

2.1. Rational $K(\pi, 1)$'s: basic examples

An important role is also played by complex **Rational** $K(\pi, 1)$**'s**, *i.e.*, quasi projective varieties (or complex spaces) Z such that

$$Z = \mathcal{D}/\pi,$$

where $\mathcal{D}$ is a contractible manifold (or complex space) and the action of π on $\mathcal{D}$ is properly discontinuous but not necessarily free.

While for a $K(\pi, 1)$ we have $H^*(\pi, \mathbb{Z}) \cong H^*(Z, \mathbb{Z})$, $H_*(\pi, \mathbb{Z}) \cong H_*(Z, \mathbb{Z})$, for a rational $K(\pi, 1)$ we have $H^*(\pi, \mathbb{Q}) \cong H^*(Z, \mathbb{Q})$ and therefore also $H_*(\pi, \mathbb{Q}) \cong H_*(Z, \mathbb{Q})$.

Typical examples of such rational $K(\pi, 1)$'s are:

(1) quotients of a bounded symmetric domain $\mathcal{D}$ by a subgroup $\Gamma \subset \mathrm{Aut}(\mathcal{D})$ which is acting properly discontinuously (equivalently, Γ is discrete); especially noteworthy are the case where Γ is **cocompact**, meaning that $X = \mathcal{D}/\Gamma$ is compact, and the **finite volume** case where the volume of X via the invariant volume form for $\mathcal{D}$ is finite.

(2) the moduli space of principally polarized Abelian Varieties, where $\mathcal{D}$ is Siegel's upper half space

$$\mathcal{H}_g := \{\tau \in \mathrm{Mat}(g, g, \mathbb{C}) | \tau = {}^t\tau, \mathrm{Im}(\tau) > 0\},$$

and the group Γ is

$$\mathrm{Sp}(2g, \mathbb{Z}) := \{M \in \mathrm{Mat}(2g, \mathbb{Z}) | {}^t MIM = I\}.$$

(3) The moduli space of curves of genus $g \geq 2$, a quotient

$$(**)\ \mathfrak{M}_g = \mathcal{T}_g / \mathcal{M}ap_g$$

of a connected complex manifold $\mathcal{T}_g$ of dimension $3g - 3$, called **Teichmüller space**, by the properly discontinuous action of the **Mapping class group** $\mathcal{M}ap_g$. A key result (see [62,66,90]) is that Teichmüller space $\mathcal{T}_g$ is diffeomorphic to a ball, and the action of $\mathcal{M}ap_g$ is properly discontinuous.

Denoting as usual by π_g the fundamental group of a compact complex curve C of genus g, we have in fact a more concrete description of the mapping class group:

$$(M)\ \mathcal{M}ap_g \cong \mathrm{Out}^+(\pi_g).$$

The above superscript $^+$ refers to the orientation preserving property.

The above isomorphism (M) is of course related to the fact that C is a $K(\pi_g, 1)$, as soon as $g \geq 1$.

As we already discussed, there is a bijection between homotopy classes of self maps of C and endomorphisms of π_g, taken up to inner conjugation. Clearly a homeomorphism $\varphi : C \to C$ yields then an associated element $\pi_1(\varphi) \in \mathrm{Out}(\pi_g)$.

Teichmüller theory can be further applied in order to analyse the fixed loci of finite subgroups G of the mapping class group (see [29,66,90]).

Theorem 2.1 (Refined Nielsen realization). *Let $G \subset \mathcal{M}ap_g$ be a finite subgroup. Then $Fix(G) \subset \mathcal{T}_g$ is a non empty complex manifold, diffeomorphic to a ball. It describes the curves which admit a group of automorphisms isomorphic to G and with a given topological action.*

3. Regularity of classifying maps and fundamental groups of projective varieties

3.1. Harmonic maps

Given a continuous map $f : M \to N$ of differentiable manifolds, we can approximate it by a differentiable one, homotopic to the previous one. Indeed, we may assume that $N \subset \mathbb{R}^n$, $M \subset \mathbb{R}^m$ and, by a partition of unity argument, that M is an open set in $\mathbb{R}^h$. Convolution approximates then f by a differentiable function F_1 with values in a tubular neighbourhood $T(N)$ of N, and then the implicit function theorem applied to the normal bundle provides a differentiable retraction $r : T(N) \to N$. Then $F := r \circ F_1$ is the required approximation, and the same retraction provides a homotopy between f and F (the homotopy between f and F_1 being obvious).

If however M, N are algebraic varieties, and algebraic topology tells us about the existence of a continuous map f as above, we would wish for more regularity, possibly holomorphicity of the homotopic map F.

Now, Wirtinger's theorem characterises complex submanifolds as area minimizing ones, so the first idea is to try to deform a differentiable mapping f until it minimizes some functional.

We may take the Riemannian structure inherited form the chosen embedding, and assume that (M, g_M), (N, g_N) are Riemannian manifolds.

If we assume that M is compact, then one defines the **Energy** $\mathcal{E}(f)$ of the map as the integral:

$$\mathcal{E}(f) := 1/2 \int_M |Df|^2 d\mu_M,$$

where Df is the derivative of the differentiable map f, $d\mu_M$ is the volume element on M, and $|Df|$ is just its norm as a differentiable section of a bundle endowed with a metric:

$$Df \in H^0(M, \mathcal{C}^\infty(TM^\vee \otimes f^*(TN))).$$

These notions were introduced by Eells and Sampson in the seminal paper [50], which used the **heat flow**

$$\frac{\partial f_t}{\partial t} = \Delta(f)$$

in order to find extremals for the energy functional. These curves in the space of maps are (as explained in [50]) the analogue of gradient lines in Morse theory, and the energy functional decreases on these lines.

The obvious advantage of the flow method with respect to discrete convergence procedures ('direct methods of the calculus of variations') is that here it is clear that all the maps are homotopic to each other! [3]

The next theorem is one of the most important results, first obtained in [50]

Theorem 3.1 (Eells-Sampson). *Let M,N be compact Riemannian manifolds, and assume that the sectional curvature $\mathcal{K}_N$ of N is semi-negative ($\mathcal{K}_N \leq 0$): then every continuous map $f_0 : M \to N$ is homotopic to a harmonic map $f : M \to N$. Moreover the equation $\Delta(f) = 0$ implies, in case where M, N are real analytic manifolds, the real analyticity of f.*

Not only the condition about the curvature is necessary for the existence of a harmonic representative in each homotopy class, but moreover it constitutes the main source of connections with the concept of classifying spaces, in view of the classical (see [71], [23]) theorem of Cartan-Hadamard establishing a deep link between curvature and topology.

Theorem 3.2 (Cartan-Hadamard). *Suppose that N is a complete Riemannian manifold, with semi-negative ($\mathcal{K}_N \leq 0$) sectional curvature: then the universal covering $\tilde{N}$ is diffeomorphic to an Euclidean space, more precisely given any two points there is a unique geodesic joining them.*

In complex dimension 1 one cannot hope for a stronger result, to have a holomorphic map rather than just a harmonic one. The surprise comes from the fact that, with suitable assumptions, the hope can be realized

[3] The flow method made then its way further through the work of Hamilton, Perelman and others, leading to the solution of the three dimensional Poincaré conjecture (see for example [75] for an exposition).

in higher dimensions, with a small proviso: given a complex manifold X, one can define the conjugate manifold $\bar{X}$ as the same differentiable manifold, but where in the decomposition $TX \otimes_{\mathbb{R}} \mathbb{C} = T^{(1,0)} \oplus T^{(0,1)}$ the roles of $T^{(1,0)}$ and $T^{(0,1)}$ are interchanged (this amounts, in case where X is an algebraic variety, to replacing the defining polynomial equations by polynomials obtained from the previous ones by applying complex conjugation to the coefficients, *i.e.*, replacing each $P(x_0, \ldots, x_N)$ by $\overline{P(\overline{x_0}, \ldots, \overline{x_N})}$).

In this case the identity map, viewed as a map $\iota : X \to \bar{X}$ is no longer holomorphic, but antiholomorphic. Assume now that we have a harmonic map $f : Y \to X$: then also $\iota \circ f$ shall be harmonic, but a theorem implying that f must be holomorphic then necessarily implies that there is a complex isomorphism between X and $\bar{X}$. Unfortunately, this is not the case, as one sees, already in the case of elliptic curves; but then one may restrict the hope to proving that f is either holomorphic or antiholomorphic.

A breakthrough in this direction was obtained by Siu ([84]) who proved several results, that we shall discuss in the next sections.

3.2. Kähler manifolds and some archetypal theorem

The assumption that a complex manifold X is a Kähler manifold is that there exists a Hermitian metric on the tangent bundle $T^{(1,0)}$ whose associated $(1,1)$ form ξ is closed. In local coordinates the metric is given by

$$h = \Sigma_{i,j} g_{i,j} dz_i d\bar{z_j}, \text{ with } d\xi = 0, \xi := (\Sigma_{i,j} g_{i,j} dz_i \wedge d\bar{z_j}).$$

Hodge theory shows that the cohomology of a compact Kähler manifold X has a Hodge-Kähler decomposition, where $H^{p,q}$ is the space of harmonic forms of type (p,q), which are in particular d-closed (and d^*-closed):

$$H^m(X, \mathbb{C}) = \oplus_{p,q \geq 0, p+q=m} H^{p,q}, \; H^{q,p} = \overline{H^{p,q}}, \, H^{p,q} \cong H^q(X, \Omega^p_X).$$

We give just an elementary application of the above theorem, a characterization of complex tori (see [28,32,35] for other characterizations)

Theorem 3.3. *Let X be a cKM,* i.e., *a compact Kähler manifold X, of dimension n. Then X is a complex torus if and only if it has the same integral cohomology algebra of a complex torus,* i.e.: $H^*(X, \mathbb{Z}) \cong \wedge^* H^1(X, \mathbb{Z})$. *Equivalently, if and only if $H^*(X, \mathbb{C}) \cong \wedge^* H^1(X, \mathbb{C})$ and $H^{2n}(X, \mathbb{Z}) \cong \wedge^{2n} H^1(X, \mathbb{Z})$*

Proof. Since $H^{2n}(X, \mathbb{Z}) \cong \mathbb{Z}$, it follows that $H^1(X, \mathbb{Z})$ is free of rank equal to $2n$, therefore $dim_{\mathbb{C}}(H^{1,0}) = n$. We consider then, chosen a base point $x_0 \in X$, the Albanese map

$$a_X : X \to \mathrm{Alb}(X) := H^0(\Omega^1_X)^\vee / \mathrm{Hom}(H^1(X, \mathbb{Z}), \mathbb{Z}),\ x \mapsto \int_{x_0}^x .$$

Therefore we have a map between X and the complex torus $T := \mathrm{Alb}(X)$, which induces an isomorphism of first cohomology groups, and has degree 1, in view of the isomorphism

$$H^{2n}(X, \mathbb{Z}) \cong \Lambda^{2n}(H^1(X, \mathbb{Z})) \cong H^{2n}(T, \mathbb{Z}).$$

In view of the normality of X, it suffices to show that a_X is finite. Let Y be a subvariety of X of dimension $m > 0$ mapping to a point: then the cohomology (or homology class, in view of Poincaré duality) class of Y is trivial, since the cohomology algebra of X and T are isomorphic. But since X is Kähler, if ξ is the Kähler form, $\int_Y \xi^m > 0$, a contradiction, since this integral depends only (by the closedness of ξ) on the homology class of Y. □

One can conjecture that a stronger theorem holds, namely

Conjecture 3.4. Let X be a cKM, *i.e.*, a compact Kähler manifold X, of dimension n. Then X is a complex torus if and only if it has the same rational cohomology algebra of a complex torus, *i.e.* $H^*(X, \mathbb{Q}) \cong \wedge^* H^1(X, \mathbb{Q})$. Equivalently, if and only if $H^*(X, \mathbb{C}) \cong \wedge^* H^1(X, \mathbb{C})$.

3.3. Siu's results on harmonic maps

The result by Siu that is the simplest to state is the following

Theorem 3.5.

(I) *Assume that $f : M \to N$ is a harmonic map between two compact Kähler manifolds and that the curvature tensor of N is strongly negative. Assume further that the real rank of the derivative Df is at least 4 in some point of M. Then f is either holomorphic or antiholomorphic.*

(II) *In particular, if $\dim_{\mathbb{C}}(N) \geq 2$ and M is homotopy equivalent to N, then M is either biholomorphic or antibiholomorphic to N.*

Let us try however to describe precisely the main hypothesis of strong negativity of the curvature, which is a stronger condition than the strict negativity of the sectional curvature.

As we already mentioned, the assumption that N is a Kähler manifold is that there exists a Hermitian metric on the tangent bundle $T^{(1,0)}$ whose associated $(1, 1)$ form is closed. In local coordinates the metric is given by

$$\Sigma_{i,j} g_{i,j} dz_i d\bar{z_j}, \quad \text{with } d(\Sigma_{i,j} g_{i,j} dz_i \wedge d\bar{z_j}) = 0.$$

The curvature tensor is a $(1, 1)$ form with values in $(T^{(1,0)})^\vee \otimes T^{(1,0)}$, and using the Hermitian metric to identify $(T^{(1,0)})^\vee \cong \overline{T^{(1,0)}} = T^{(0,1)}$, and their conjugates $((T^{(0,1)})^\vee = \overline{(T^{(0,1)})} \cong T^{(1,0)}$) we write as usual the curvature tensor as a section R of

$$(T^{(1,0)})^\vee \otimes (T^{(0,1)})^\vee \otimes (T^{(1,0)})^\vee \otimes (T^{(0,1)})^\vee.$$

Then seminegativity of the sectional curvature is equivalent to

$$-R(\xi \wedge \bar{\eta} - \eta \wedge \bar{\xi}, \overline{\xi \wedge \bar{\eta} - \eta \wedge \bar{\xi}}) \leq 0,$$

for all pairs of complex tangent vectors ξ, η (here one uses the isomorphism $T^{(1,0)} \cong TN$, and one sees that the expression depends only on the real span of the two vectors ξ, η).

Strong negativity means instead that

$$-R(\xi \wedge \bar{\eta} - \zeta \wedge \bar{\theta}, \overline{\xi \wedge \bar{\eta} - \zeta \wedge \bar{\theta}}) < 0,$$

for all 4-tuples of complex tangent vectors ξ, η, ζ, θ.

The geometrical meaning is the following (see [1, page 71]): the sectional curvature is a quadratic form on $\wedge^2(TN)$, and as such it extends to the complexified bundle $\wedge^2(TN) \otimes \mathbb{C}$ as a Hermitian form. Then strong negativity in the sense of Siu is also called negativity of the **Hermitian sectional curvature** $R(v, w, \bar{v}, \bar{w})$ for all vectors $v, w \in (TN) \otimes \mathbb{C}$.

Then a reformulation of the result of Siu ([84]) and Sampson ([81]) is the following:

Theorem 3.6. *Let M be a compact Kähler manifold, and N a Riemannian manifold with semi-negative Hermitian sectional curvature. Then every harmonic map $f : M \to N$ is pluri-harmonic.*

Now, examples of varieties N with a strongly negative curvature are the balls in $\mathbb{C}^n$, *i.e.*, the BSD of type $I_{n,1}$; Siu finds out that ([84]) for the irreducible bounded symmetric domains of type

$$I_{p,q}, \text{ for } pq \geq 2, \ II_n, \forall n \geq 3, III_n, \forall n \geq 2, IV_n, \forall n \geq 3,$$

the metric is not strongly negative, but just very strongly seminegative, where very strong negativity simply means negativity of the curvature as a Hermitian form on $T^{1,0} \otimes T^{0,1} = T^{1,0} \otimes \overline{T^{0,1}}$. This gives rise to several technical difficulties, where the bulk of the calculations is to see that there is an upper bound for the nullity of the Hermitian sectional curvature, *i.e.* for the rank of the real subbundles of TM where the Hermitian sectional curvature restricts identically to zero.

Siu derives then several results, and we refer the reader to the book [1] for a nice exposition of these results of Siu.

3.4. Hodge theory and existence of maps to curves

Siu also used harmonic theory in order to construct holomorphic maps from Kähler manifolds to projective curves. This the theorem of [86]

Theorem 3.7 (Siu). *Assume that a compact Kähler manifold X is such that there is a surjection $\phi : \pi_1(X) \to \pi_g$, where $g \geq 2$ and, as usual, π_g is the fundamental groups of a projective curve of genus g. Then there is a projective curve C of genus $g' \geq g$ and a fibration $f : X \to C$ (*i.e.*, the fibres of f are connected) such that ϕ factors through $\pi_1(f)$.*

In this case the homomorphism leads to a harmonic map to a curve, and one has to show that the Stein factorization yields a map to some Riemann surface which is holomorphic for some complex structure on the target.

In this case it can be seen more directly how the Kähler assumption, which boils down to Kähler identities, is used.

Recall that Hodge theory shows that the cohomology of a compact Kähler manifold X has a Hodge-Kähler decomposition, where $H^{p,q}$ is the space of harmonic forms of type (p, q):

$$H^m(X, \mathbb{C}) = \oplus_{p,q\geq 0, p+q=m} H^{p,q}, \quad H^{q,p} = \overline{H^{p,q}}, \, H^{p,q} \cong H^q(X, \Omega^p_X).$$

The Hodge-Kähler decomposition theorem has a long story, and was proven by Picard in special cases. It entails the following consequence:

Holomorphic forms are closed, *i.e.*, $\eta \in H^0(X, \Omega^p_X) \Rightarrow d\eta = 0$.

At the turn of last century this fact was then used by Castelnuovo and de Franchis ([25,46]):

Theorem 3.8 (Castelnuovo-de Franchis). *Assume that X is a compact Kähler manifold, $\eta_1, \eta_2 \in H^0(X, \Omega^1_X)$ are $\mathbb{C}$-linearly independent, and the wedge product $\eta_1 \wedge \eta_2$ is d-exact. Then $\eta_1 \wedge \eta_2 \equiv 0$ and there exists a fibration $f : X \to C$ such that $\eta_1, \eta_2 \in f^* H^0(C, \Omega^1_C)$. In particular, C has genus $g \geq 2$.*

From it one gets the following simple theorem ([27]):

Theorem 3.9 (Isotropic subspace theorem). *On a compact Kähler manifold X there is a bijection between isomorphism classes of fibrations $f : X \to C$ to a projective curve of genus $g \geq 2$, and real subspaces $V \subset H^1(X, \mathbb{C})$ ('real' means that V is self conjugate, $\overline{V} = V$) which have dimension 2g and are of the form $V = U \oplus \bar{U}$, where U is a maximal isotropic subspace for the wedge product*

$$H^1(X, \mathbb{C}) \times H^1(X, \mathbb{C}) \to H^2(X, \mathbb{C}).$$

Another result in this vein is (*cf.* [34]).

Theorem 3.10. *Let X be a compact Kähler manifold, and let $f : X \to C$ be a fibration onto a projective curve C, of genus g, and assume that there are exactly r fibres which are multiple with multiplicities $m_1, \dots m_r \geq 2$. Then f induces an orbifold fundamental group exact sequence*

$$\pi_1(F) \to \pi_1(X) \to \pi_1(g; m_1, \dots m_r) \to 0,$$

where F is a smooth fibre of f, and

$$\pi_1(g; m_1, \dots m_r) = \langle \alpha_1, \beta_1, \dots, \alpha_g, \beta_g, \gamma_1, \dots \gamma_r | \Pi_1^g [\alpha_j, \beta_j] \Pi_1^r \gamma_i = \gamma_1^{m_1} = \dots = \gamma_r^{m_r} = 1 \rangle.$$

Conversely, let X be a compact Kähler manifold and let $(g, m_1, \dots m_r)$ be a hyperbolic type, i.e., *assume that $2g - 2 + \Sigma_i (1 - \frac{1}{m_i}) > 0$.*

Then each epimorphism $\phi : \pi_1(X) \to \pi_1(g; m_1, \dots m_r)$ with finitely generated kernel is obtained from a fibration $f : X \to C$ of type $(g; m_1, \dots m_r)$.

The following (see [29] and [30]) is the main result concerning surfaces isogenous to a product.

Theorem 3.11.

a) *A projective smooth surface S is isogenous to a product of two curves of respective genera $g_1, g_2 \geq 2$, if and only if the following two conditions are satisfied:*

1) *there is an exact sequence*

$$1 \to \pi_{g_1} \times \pi_{g_2} \to \pi = \pi_1(S) \to G \to 1,$$

where G is a finite group and where π_{g_i} denotes the fundamental group of a projective curve of genus $g_i \geq 2$;

2) $e(S) (= c_2(S)) = \frac{4}{|G|}(g_1 - 1)(g_2 - 1)$.

b) *Write $S = (C_1 \times C_2)/G$. Any surface X with the same topological Euler number and the same fundamental group as S is diffeomorphic to S and is also isogenous to a product. There is a smooth proper family with connected smooth base manifold T, $p : \mathcal{X} \to T$ having two fibres respectively isomorphic to X, and Y, where Y is one of the 4 surfaces $S = (C_1 \times C_2)/G$, $S_{+-} := (\overline{C_1} \times C_2)/G$, $\bar{S} = (\overline{C_1} \times \overline{C_2})/G$, $S_{-+} := (C_1 \times \overline{C_2})/G = \overline{S_{+-}}$.*

c) *The corresponding subset of the moduli space of surfaces of general type $\mathfrak{M}_S^{top} = \mathfrak{M}_S^{diff}$, corresponding to surfaces orientedly homeomorphic, resp. orientedly diffeomorphic to S, is either irreducible and connected or it contains two connected components which are exchanged by complex conjugation.*
In particular, if S' is orientedly diffeomorphic to S, then S' is deformation equivalent to S or to $\bar{S}$.

4. Inoue type varieties

While a couple of hundreds examples are known today of families of minimal surfaces of general type with geometric genus $p_g(S) := \dim H^0(\mathcal{O}_S(K_S)) = 0$ (observe that for these surfaces $1 \leq K_S^2 \leq 9$), for the value $K_S^2 = 7$ there are only two examples known (*cf.* [63] and [44]), and for a long time only one family of such surfaces was known, the one constructed by Masahisa Inoue (*cf.* [63]).

The attempt to prove that Inoue surfaces form a connected component of the moduli space of surfaces of general type proved to be successful ([15]), and was based on a weak rigidity result: the topological type of an Inoue surface determines an irreducible connected component of the moduli space (a phenomenon similar to the one which was observed in several papers, as [10,11,16,43]).

The starting point was the calculation of the fundamental group of an Inoue surface with $p_g = 0$ and $K_S^2 = 7$: it sits in an extension (π_g being as usual the fundamental group of a projective curve of genus g):

$$1 \to \pi_5 \times \mathbb{Z}^4 \to \pi_1(S) \to (\mathbb{Z}/2\mathbb{Z})^5 \to 1.$$

This extension is given geometrically, *i.e.*, stems from the observation ([15]) that an Inoue surface S admits an unramified $(\mathbb{Z}/2\mathbb{Z})^5$ - Galois covering $\hat{S}$ which is an ample divisor in $E_1 \times E_2 \times D$, where E_1, E_2 are elliptic curves and D is a projective curve of genus 5; while Inoue described $\hat{S}$ as a complete intersection of two non ample divisors in the product $E_1 \times E_2 \times E_3 \times E_4$ of four elliptic curves.

It turned out that the ideas needed to treat this special family of Inoue surfaces could be put in a rather general framework, valid in all dimensions, setting then the stage for the investigation and search for a new class of varieties, which we proposed to call Inoue-type varieties.

Definition 4.1 ([15]). Define a complex projective manifold X to be an **Inoue-type manifold** if

(1) $\dim(X) \geq 2$;

(2) there is a finite group G and an unramified G-covering $\hat{X} \to X$, (hence $X = \hat{X}/G$) such that

(3) $\hat{X}$ is an ample divisor inside a $K(\Gamma, 1)$-projective manifold Z, (hence by the theorems of Lefschetz, $\pi_1(\hat{X}) \cong \pi_1(Z) \cong \Gamma$) and moreover

(4) the action of G on $\hat{X}$ yields a faithful action on $\pi_1(\hat{X}) \cong \Gamma$: in other words the exact sequence

$$1 \to \Gamma \cong \pi_1(\hat{X}) \to \pi_1(X) \to G \to 1$$

gives an injection $G \to \mathrm{Out}(\Gamma)$, induced by conjugation by lifts of elements of G.

(5) the action of G on $\hat{X}$ is induced by an action of G on Z.

Similarly one defines the notion of an **Inoue-type variety**, by requiring the same properties for a variety X with canonical singularities.

The above definition of Inoue type manifold, although imposing a strong restriction on X, is too general, and in order to get weak rigidity type results it is convenient to impose restrictions on the fundamental group Γ of Z, for instance the most interesting case is the one where Z is a product of Abelian varieties, curves, and other locally symmetric varieties with ample canonical bundle.

Definition 4.2. We shall say that an Inoue-type manifold X is

(1) a **special Inoue type manifold** if moreover

$$Z = (A_1 \times \cdots \times A_r) \times (C_1 \times \cdots \times C_h) \times (M_1 \times \cdots \times M_s)$$

where each A_i is an Abelian variety, each C_j is a curve of genus $g_j \geq 2$, and M_i is a compact quotient of an irreducible bounded symmetric domain of dimension at least 2 by a torsion free subgroup;

(2) a **classical Inoue type manifold** if moreover $Z = (A_1 \times \cdots \times A_r) \times (C_1 \times \cdots \times C_h)$ where each A_i is an Abelian variety, each C_j is a curve of genus $g_j \geq 2$;

(3) a special Inoue type manifold is said to be **diagonal** if moreover:

(I) the action of G on $\hat{X}$ is induced by a diagonal action on Z, *i.e.*,

$$G \subset \prod_{i=1}^{r} \mathrm{Aut}(A_i) \times \prod_{j=1}^{h} \mathrm{Aut}(C_j) \times \prod_{l=1}^{s} \mathrm{Aut}(M_l) \tag{4.1}$$

and furthermore:

(II) the faithful action on $\pi_1(\hat{X}) \cong \Gamma$, induced by conjugation by lifts of elements of G in the exact sequence

$$\begin{aligned} 1 \to \Gamma = \Pi_{i=1}^{r}(\Lambda_i) \times \Pi_{j=1}^{h}(\pi_{g_j}) \times \Pi_{l=1}^{s}(\pi_1(M_l)) \\ \to \pi_1(X) \to G \to 1 \end{aligned} \tag{4.2}$$

(observe that each factor Λ_i, resp. π_{g_j}, $\pi_1(M_l)$ is a normal subgroup), satisfies the **Schur property**

$$(SP) \quad \mathrm{Hom}(V_i, V_j)^G = 0, \forall i \neq j.$$

Here $V_j := \Lambda_j \otimes \mathbb{Q}$ and, in order that the Schur property holds, it suffices for instance to verify that for each Λ_i there is a subgroup H_i of G for which $\mathrm{Hom}(V_i, V_j)^{H_i} = 0, \forall j \neq i$.

The Schur property (SP) plays an important role in order to show that an Abelian variety with such a G-action on its fundamental group must split as a product.

Before stating the main general result of [15] we need the following definition, which was already used in 3.3 for the characterization of complex tori among Kähler manifolds.

Definition 4.3. Let Y, Y' be two projective manifolds with isomorphic fundamental groups. We identify the respective fundamental groups $\pi_1(Y) = \pi_1(Y') = \Gamma$. Then we say that the condition **(SAME HOMOLOGY)** is satisfied for Y and Y' if there is an isomorphism $\Psi : H_*(Y', \mathbb{Z}) \cong H_*(Y, \mathbb{Z})$ of homology groups which is compatible with the homomorphisms

$$u\colon H_*(Y, \mathbb{Z}) \to H_*(\Gamma, \mathbb{Z}),\, u'\colon H_*(Y', \mathbb{Z}) \to H_*(\Gamma, \mathbb{Z}),$$

i.e., Ψ satisfies $u \circ \Psi = u'$.

We can now state the following

Theorem 4.4. *Let X be a diagonal special Inoue type manifold, and let X' be a projective manifold with $K_{X'}$ nef and with the same fundamental group as X, which moreover either*

(A) *is homotopically equivalent to* X*;*
or satisfies the following weaker property:
(B) *let* $\hat{X}'$ *be the corresponding unramified covering of* X'*. Then* $\hat{X}$ *and* $\hat{X}'$ *satisfy the condition* **(SAME HOMOLOGY)**.
Setting $W := \hat{X}'$*, we have that*

   (a) $X' = W/G$ *where* W *admits a generically finite morphism* $f : W \to Z'$*, and where* Z' *is also a* $K(\Gamma, 1)$ *projective manifold, of the form* $Z' = (A'_1 \times \cdots \times A'_r) \times (C'_1 \times \cdots \times C'_h) \times (M'_1 \times \cdots \times M'_s)$*. Moreover here* M'_i *is either* M_i *or its complex conjugate, and the product decomposition corresponds to the product decomposition* (4.2) *of the fundamental group of* Z*.*
   The image cohomology class $f_*([W])$ *corresponds, up to sign, to the cohomology class of* $\hat{X}$*.*
   (b) *The morphism* f *is finite if* $n = \dim X$ *is odd, and it is generically injective if*
   (**) *the cohomology class of* $\hat{X}$ *(in* $H^*(Z, \mathbb{Z})$*) is indivisible, or if every strictly submultiple cohomology class cannot be represented by an effective* G*-invariant divisor on any pair* (Z', G) *homotopically equivalent to* (Z, G)*.*
   (c) f *is an embedding if moreover* $K_{X'}$ *is ample,*
   (*) *every such divisor* W *of* Z' *is ample, and*
   (***) $K^n_{X'} = K^n_X$*.*[4]

In particular, if $K_{X'}$ *is ample and* (*), (**) *and* (***) *hold, also* X' *is a diagonal SIT (special Inoue type) manifold.*
A similar conclusion holds under the alternative assumption that the homotopy equivalence sends the canonical class of W *to that of* $\hat{X}$*: then* X' *is a minimal resolution of a diagonal SIT (special Inoue type) variety.*

For the proof of Theorem 4.4 the first step consists in showing that $W := \hat{X}'$ admits a holomorphic mapping to a manifold Z' of the above type $Z' = (A'_1 \times \cdots \times A'_r) \times (C'_1 \times \cdots \times C'_h) \times (M'_1 \times \cdots \times M'_s)$, where M'_i is either M_i or its complex conjugate.

First of all, by the results of Siu and others ([31, 84, 85], [34, Theorem 5.14]) cited in Section 3, W admits a holomorphic map to a product manifold of the desired type

$$Z'_2 \times Z'_3 = (C'_1 \times \cdots \times C'_h) \times (M'_1 \times \cdots \times M'_s).$$

[4] This last property for algebraic surfaces follows automatically from homotopy invariance.

Then one looks at the Albanese variety Alb(W) of the Kähler manifold W, whose fundamental group is the quotient of the Abelianization of $\Gamma = \pi_1(Z)$ by its torsion subgroup.

Then the cohomological assumptions and adjunction theory are used to complete the result.

The study of moduli spaces of Inoue type varieties, and their connected and irreducible components, relies very much on the study of moduli spaces of varieties X endowed with the action of a finite group G: and it is for us a strong motivation to pursue this line of research.

This topic will occupy a central role in the following sections, first in general, and then in the special case of algebraic curves.

5. Moduli spaces of symmetry marked varieties

5.1. Moduli marked varieties

We give now the definition of a symmetry marked variety for projective varieties, but one can similarly give the same definition for complex or Kähler manifolds; to understand the concept of a marking, it suffices to consider a cyclic group acting on a variety X. A marking consists in this case of the choice of a generator for the group acting on X. The marking is very important when we have several actions of a group G on some projective varieties, and we want to consider the diagonal action of G on their product.

Definition 5.1.

(1) A *G-marked (projective) variety* is a triple (X, G, η) where X is a projective variety, G is a group and $\eta\colon G \to \mathrm{Aut}(X)$ is an injective homomorphism

(2) equivalently, a marked variety is a triple (X, G, α) where $\alpha\colon X \times G \to X$ is a faithful action of the group G on X.

(3) Two marked varieties (X, G, α), (X', G, α') are said to be *isomorphic* if there is an isomorphism $f\colon X \to X'$ transporting the action $\alpha\colon X \times G \to X$ into the action $\alpha'\colon X' \times G \to X'$, *i.e.*, such that

$$f \circ \alpha = \alpha' \circ (f \times \mathrm{id}) \Leftrightarrow \eta' = Ad(f) \circ \eta, \quad Ad(f)(\phi) := f\phi f^{-1}.$$

(4) If G is a subset of Aut(X), then the natural marked variety is the triple (X, G, i), where $i\colon G \to \mathrm{Aut}(X)$ is the inclusion map, and it shall sometimes be denoted simply by the pair (X, G).

(5) A marked curve (D, G, η) consisting of a smooth projective curve of genus g and a faithful action of the group G on D is said to be a *marked triangle curve of genus g* if $D/G \cong \mathbb{P}^1$ and the quotient morphism $p\colon D \to D/G \cong \mathbb{P}^1$ is branched in three points.

Remark 5.2. Observe that:

1) we have a natural action of Aut(G) on G-marked varieties, namely, if $\psi \in \mathrm{Aut}(G)$,
$$\psi(X, G, \eta) := (X, G, \eta \circ \psi^{-1}).$$
The corresponding equivalence class of a G-marked variety is defined to be a *G-(unmarked) variety*.
2) the action of the group Inn(G) of inner automorphisms does not change the isomorphism class of (X, G, η) since, for $\gamma \in G$, we may set $f := \eta(\gamma)$, $\psi := Ad(\gamma)$, and then $\eta \circ \psi = Ad(f) \circ \eta$, since $\eta(\psi(g)) = \eta(\gamma g \gamma^{-1}) = \eta(\gamma)\eta(g)(\eta(\gamma))^{-1}) = Ad(f)(\eta(g))$.
3) In the case where $G = \mathrm{Aut}(X)$, we see that Out(G) acts simply transitively on the isomorphism classes of the Aut(G)-orbit of (X, G, η).

Let us see now how the picture works in the case of curves: this case is already very enlightening and intriguing.

5.2. Moduli of curves with automorphisms

There are several 'moduli spaces' of curves with automorphisms. First of all, given a finite group G, we define a subset $\mathfrak{M}_{g,G}$ of the moduli space $\mathfrak{M}_g$ of smooth curves of genus $g > 1$: $\mathfrak{M}_{g,G}$ is the locus of the curves that admit an effective action by the group G. It turns out that $\mathfrak{M}_{g,G}$ is a Zariski closed algebraic subset. The description of these Zariski closed subsets is related to the description of the singular locus of the moduli space $\mathfrak{M}_g$ (for instance of its irreducible components, see [45]), and of its compactification $\overline{\mathfrak{M}_g}$, (see [36]).

In order to understand the irreducible components of $\mathfrak{M}_{g,G}$ we have seen that Teichmüller theory plays an important role: it shows the connectedness, given an injective homomorphism $\rho\colon G \to \mathrm{Map}_g$, of the locus
$$\mathcal{T}_{g,\rho} := Fix(\rho(G)).$$
Its image $\mathfrak{M}_{g,\rho}$ in $\mathfrak{M}_{g,G}$ is a Zariski closed irreducible subset (as observed in [39]). Recall that to a curve C of genus g with an action by G we can associate several discrete invariants that are constant under deformation.

The first is the above *topological type* of the G-action: it is a homomorphism $\rho\colon G \to \mathrm{Map}_g$, which is well-defined up to inner conjugation (induced by different choices of an isomorphism $\mathrm{Map}(C) \cong \mathrm{Map}_g$).

We immediately see that the locus $\mathfrak{M}_{g,\rho}$ is first of all determined by the subgroup $\rho(G)$ and not by the marking. Moreover, this locus remains the same not only if we change ρ modulo the action by Aut(G), but also if we change ρ by the adjoint action by Map_g.

Definition 5.3.

1) The moduli space of G-marked curves of a certain topological type ρ is the quotient of the Teichmüller submanifold $\mathcal{T}_{g,\rho}$ by the centralizer subgroup $\mathcal{C}_{\rho(G)}$ of the subgroup $\rho(G)$ of the mapping class group. We get a normal complex space which we shall denote $\mathfrak{M}_g[\rho]$. $\mathfrak{M}_g[\rho] = \mathcal{T}_{g,\rho}/\mathcal{C}_{\rho(G)}$ is a finite covering of a Zariski closed subset of the usual moduli space (its image $\mathfrak{M}_{g,\rho}$), therefore it is quasi-projective, by the theorem of Grauert and Remmert.
2) Defining $\mathfrak{M}_g(\rho)$ as the quotient of $\mathcal{T}_{g,\rho}$ by the normalizer $\mathcal{N}_{\rho(G)}$ of $\rho(G)$, we call it the moduli space of curves with a G-action of a given topological type. It is again a normal quasi-projective variety.

Remark 5.4.

1) If we consider $G' := \rho(G)$ as a subgroup $G' \subset \mathrm{Map}_g$, then we get a natural G'-marking for any $C \in Fix(G') = \mathcal{T}_{g,\rho}$.
2) As we said, $Fix(G') = \mathcal{T}_{g,\rho}$ is independent of the chosen marking, moreover the projection $Fix(G') = \mathcal{T}_{g,\rho} \to \mathfrak{M}_{g,\rho}$ factors through a finite map $\mathfrak{M}_g(\rho) \to \mathfrak{M}_{g,\rho}$.

The next question is whether $\mathfrak{M}_g(\rho)$ maps 1-1 into the moduli space of curves. This is not the case, as we shall easily see. Hence one gives the following definition.

Definition 5.5. Let $G \subset \mathrm{Map}_g$ be a finite group, and let C represent a point in $Fix(G)$. Then we have a natural inclusion $G \to A_C := \mathrm{Aut}(C)$, and C is a fixed point for the subgroup $A_C \subset \mathrm{Map}_g$: A_C is indeed the stabilizer of the point C in Map_g, so that locally (at the point of $\mathfrak{M}_g$ corresponding to C) we get a complex analytic isomorphism $\mathfrak{M}_g = \mathcal{T}_g / A_C$.

We define $H_G := \cap_{C \in Fix(G)} A_C$ and we shall say that G is a **full subgroup** if $G = H_G$. Equivalently, H_G is the largest subgroup H such that $Fix(H) = Fix(G)$.

This implies that H_G is a full subgroup.

Then we have:

Proposition 5.6. *If H is a full subgroup $H \subset \mathrm{Map}_g$, and $\rho : H \subset \mathrm{Map}_g$ is the inclusion homomorphism, then $\mathfrak{M}_g(\rho)$ is the normalization of $\mathfrak{M}_{g,\rho}$.*

5.3. Numerical and homological invariants of group actions on curves

As already mentioned, given an effective action of a finite group G on C, we set $C' := C/G$, $g' := g(C')$, and we have the quotient morphism $p: C \to C/G =: C'$, a G-cover.

The geometry of p encodes several numerical invariants that are constant on $M_{g,\rho}(G)$: first of all the genus g' of C', then the number d of branch points $y_1, \dots, y_d \in C'$.

We call the set $B = \{y_1, \dots, y_d\}$ the branch locus, and for each y_i we denote by m_i the multiplicity of y_i (the greatest number dividing the divisor $p^*(y_i)$). We choose an ordering of B such that $m_1 \leq \dots \leq m_d$.

These numerical invariants $g', d, m_1 \leq \dots \leq m_d$ form the so-called **primary numerical type**.

$p \colon C \to C'$ is determined (Riemann's existence theorem) by the monodromy, a surjective homomorphism:

$$\mu \colon \pi_1(C' \setminus B) \to G\,.$$

We have:

$$\pi_1(C' \setminus B) \cong \Pi_{g',d}$$
$$:= \langle \gamma_1, \dots, \gamma_d, \alpha_1, \beta_1, \dots, \alpha_{g'}, \beta_{g'} | \prod_{i=1}^{d} \gamma_i \prod_{j=1}^{g'} [\alpha_j, \beta_j] = 1 \rangle.$$

We set then $c_i := \mu(\gamma_i)$, $a_j := \mu(\alpha_j)$, $b_j := \mu(\beta_j)$, thus obtaining a Hurwitz generating vector, *i.e.* a vector

$$v := (c_1, \dots, c_d, a_1, b_1, \dots, a_{g'}, b_{g'}) \in G^{d+2g'}$$

s.t.

- G is generated by the entries $c_1, \dots, c_d, a_1, b_1, \dots, a_{g'}, b_{g'}$,
- $c_i \neq 1_G, \forall i$, and
- $\prod_{i=1}^{d} c_i \prod_{j=1}^{g'} [a_j, b_j] = 1$.

We see that the monodromy μ is completely equivalent, once an isomorphism $\pi_1(C' \setminus B) \cong \Pi_{g',d}$ is chosen, to the datum of a Hurwitz generating vector (we also call the sequence $c_1, \dots, c_d, a_1, b_1, \dots, a_{g'}, b_{g'}$ of the vector's coordinates a *Hurwitz generating system*).

A second numerical invariant of these components of $\mathfrak{M}_g(G)$ is obtained from the monodromy $\mu \colon \pi_1(C' \setminus \{y_1, \dots, y_d\}) \to G$ of the restriction of p to $p^{-1}(C' \setminus \{y_1, \dots, y_d\})$, and is called the ν-*type* or Nielsen function of the covering.

The Nielsen function ν is a function defined on the set of conjugacy classes in G which, for each conjugacy class $\mathcal{C}$ in G, counts the number $\nu(\mathcal{C})$ of local monodromies $c_1, \dots, c_d$ which belong to $\mathcal{C}$ (observe that the numbers $m_1 \leq \dots \leq m_d$ are just the orders of the local monodromies).

Observe in fact that the generators γ_j are well defined only up to conjugation in the group $\pi_1(C' \setminus \{y_1, \dots, y_d\})$, hence the local monodromies are well defined only up to conjugation in the group G.

We have already observed that the irreducible closed algebraic sets $M_{g,\rho}(G)$ depend only upon what we call the 'unmarked topological type', which is defined as the conjugacy class of the subgroup $\rho(G)$ inside Map_g. This concept remains however still mysterious, due to the complicated nature of the group Map_g. Therefore one tries to use more geometry to get a grasp on the topological type.

The following is immediate by Riemann's existence theorem and the irreducibility of the moduli space $\mathfrak{M}_{g',d}$ of d-pointed curves of genus g'. Given g' and d, the unmarked topological types whose primary numerical type is of the form $g', d, m_1, \dots, m_d$ are in bijection with the quotient of the set of the corresponding monodromies μ modulo the actions by $\mathrm{Aut}(G)$ and by $\mathrm{Map}(g', d)$.
Here $\mathrm{Map}(g', d)$ is the full mapping class group of genus g' and d unordered points.

Thus Riemann's existence theorem shows that the components of the moduli space

$$\mathfrak{M}(G) := \cup_g \mathfrak{M}_g(G)$$

with numerical invariants g', d correspond to the following quotient set.

Definition 5.7.

$$\mathcal{A}(g', d, G) := \mathrm{Epi}(\Pi_{g',d}, G)/\mathrm{Map}_{g',d} \times \mathrm{Aut}(G)\,.$$

Thus a first step toward the general problem consists in finding a fine invariant that distinguishes these orbits.

In the paper [39] we introduced a new homological invariant $\hat{\epsilon}$ for G-actions on smooth curves and showed that, in the case where G is the dihedral group D_n of order $2n$, $\hat{\epsilon}$ is a fine invariant since it distinguishes the different unmarked topological types.

This invariant generalizes the classical homological invariant in the unramified case.

Definition 5.8. Let $p\colon C \to C/G =: C'$ be unramified, so that $d = 0$ and we have a monodromy $\mu\colon \pi_1(C') \to G$.

Since C' is a classifying space for the group $\pi_{g'}$, we obtain a continuous map

$$m : C' \to BG,\ \pi_1(m) = \mu.$$

Moreover, $H_2(C', \mathbb{Z})$ has a natural generator $[C']$, the fundamental class of C' determined by the orientation induced by the complex structure of C'.

The homological invariant of the G-marked action is then defined as:

$$\epsilon := H_2(m)([C']) \in H_2(BG, \mathbb{Z}) = H_2(G, \mathbb{Z}).$$

If we forget the marking we have to take ϵ as an element in $H_2(G, \mathbb{Z})/\mathrm{Aut}(G)$.

In the ramified case, one needs also the following definition.

Definition 5.9. An element

$$\nu = (n_C)_C \in \bigoplus_{C \neq \{1\}} \mathbb{N}\langle C \rangle$$

is **admissible** if the following equality holds in the $\mathbb{Z}$-module G^{ab}:

$$\sum_C n_C \cdot [C] = 0$$

(here $[C]$ denotes the image element of C in the abelianization G^{ab}).

The main result of [40] is the following 'genus stabilization' theorem.

Theorem 5.10. *There is an integer h such that for $g' > h$*

$$\hat{\epsilon} \colon \mathcal{A}(g', d, G) \to (K^{\cup})/_{\mathrm{Aut}(G)}$$

induces a bijection onto the set of admissible classes of refined homology invariants.

In particular, if $g' > h$, and we have two Hurwitz generating systems v_1, v_2 having the same Nielsen function, they are equivalent if and only if the 'difference' $\hat{\epsilon}(v_1)\hat{\epsilon}(v_2)^{-1} \in H_{2,\Gamma}(G)$ is trivial.

The above result extends a nice theorem of Livingston, Dunfield and Thurston ([47, 69]) in the unramifed case, where also the statement is simpler.

Theorem 5.11. *For $g' >> 0$*

$$\hat{\epsilon} \colon \mathcal{A}(g', 0, G) \to H_2(G, \mathbb{Z})/_{\mathrm{Aut}(G)}$$

is a bijection.

Remark 5.12. Unfortunately the integer h in Theorem 5.10, which depends on the group G, is not explicit.

A key concept used in the proof is the concept of genus stabilization of a covering, which we now briefly explain.

Definition 5.13. Consider a group action of G on a projective curve C, and let $C \to C' = C/G$ the quotient morphism, with monodromy

$$\mu\colon \pi_1(C' \setminus B) \to G$$

(here B is as usual the branch locus). Then the first genus stabilization of the differentiable covering is defined geometrically by simply adding a handle to the curve C', on which the covering is trivial.

Algebraically, given the monodromy homomorphism

$$\mu : \pi_1(C' \setminus B) \cong \Pi_{g',d}$$
$$:= \left\langle \gamma_1, \dots, \gamma_d, \alpha_1, \beta_1, \dots, \alpha_{g'}, \beta_{g'} \middle| \prod_{i=1}^{d} \gamma_i \prod_{j=1}^{g'} [\alpha_j, \beta_j] = 1 \right\rangle \to G,$$

we simply extend μ to $\mu^1\colon \Pi_{g'+1,d} \to G$ setting

$$\mu^1(\alpha_{g'+1}) = \mu^1(\beta_{g'+1}) = 1_G.$$

In terms of Hurwitz vectors and Hurwitz generating systems, we replace the vector

$$v := (c_1, \dots, c_d, a_1, b_1, \dots, a_{g'}, b_{g'}) \in G^{d+2g'}$$

by

$$v^1 := (c_1, \dots, c_d, a_1, b_1, \dots, a_{g'}, b_{g'}, 1, 1) \in G^{d+2g'+2}.$$

The operation of first genus stabilization generates then an equivalence relation among monodromies (equivalently, Hurwitz generating systems), called **stable equivalence**.

The most important step in the proof, the geometric understanding of the invariant $\epsilon \in H_2(G, \mathbb{Z})$ was obtained by Livingston [69].

Theorem 5.14. *Two monodromies μ_1, μ_2 are stably equivalent if and only if they have the same invariant $\epsilon \in H_2(G, \mathbb{Z})$.*

A purely algebraic proof of Livingston's theorem was given by Zimmermann in [96], while a nice sketch of proof was given by Dunfield and Thurston in [47].

5.4. Classification results for certain concrete groups

The first result in this direction was obtained by Nielsen ([78]) who proved that ν determines ρ if G is cyclic (in fact in this case $H_2(G, \mathbb{Z}) = 0$!).

In the cyclic case the Nielsen function for $G = \mathbb{Z}/n$ is simply a function $\nu : (\mathbb{Z}/n) \setminus \{0\} \to \mathbb{N}$, and admissibility here simply means that

$$\sum_i i \cdot \nu(i) \equiv 0 \pmod n.$$

The class of ν is just the equivalence class for the equivalence relation $\nu(i) \sim \nu_r(i), \forall r \in (\mathbb{Z}/n)^*$, where $\nu_r(i) := \nu(ri), \forall i \in (\mathbb{Z}/n)$.

From the refined Nielsen realization theorem of [29] (2.1) it follows that the components of $\mathfrak{M}_g(\mathbb{Z}/n)$ are in bijection with the classes of Nielsen functions (see also [36] for an elementary proof).

The genus g' and the Nielsen class (which refine the primary numerical type), and the homological invariant $h \in H_2(G/H, \mathbb{Z})$ (here H is again the subgroup generated by the local monodromies) determine the connected components of $\mathfrak{M}_g(G)$ under some restrictions: for instance when G is abelian or when G acts freely and is the semi-direct product of two finite cyclic groups (as it follows by combining results from [29,36,48] and [49]).

Theorem 5.15 (Edmonds). *ν and $h \in H_2(G/H, \mathbb{Z})$ determine ρ for G abelian. Moreover, if G is split-metacyclic and the action is free, then h determines ρ.*

However, in general, these invariants are not enough to distinguish unmarked topological types, as one can see already for non-free D_n-actions (see [39]). Already for dihedral groups, one needs the refined homological invariant $\hat{\epsilon}$.

Theorem 5.16 ([39]). *For the dihedral group $G = D_n$ the connected components of the moduli space $\mathfrak{M}_g(D_n)$ are in bijection, via the map $\hat{\epsilon}$, with the admissible classes of refined homology invariants.*

The above result completes the classification of the unmarked topological types for $G = D_n$; moreover this result entails the classification of the irreducible components of the loci $\mathfrak{M}_{g,D_n}$ (see the appendix to [39]).

It is an interesting question: for which groups G does the refined homology invariant $\hat{\epsilon}$ determine the connected components of $\mathfrak{M}_g(G)$?

In view of Edmonds' result in the unramified case, it is reasonable to expect a positive answer for split metacyclic groups.

As mentioned in [47, page 499], the group $G = \mathbb{P}SL(2, \mathbb{F}_{13})$ shows that, for $g' = 2$, in the unramified case there are different components with trivial homology invariant $\epsilon \in H_2(G, \mathbb{Z})$: these topological types of coverings are therefore stably equivalent but not equivalent.

6. Connected components of moduli spaces and the action of the absolute Galois group

Let X be a complex projective variety: let us quickly recall the notion of a conjugate variety.

Remark 6.1.

1) $\phi \in \mathrm{Aut}(\mathbb{C})$ acts on $\mathbb{C}[z_0, \ldots z_n]$, by sending $P(z) = \sum_{i=0}^n a_i z^i \mapsto \phi(P)(z) := \sum_{i=0}^n \phi(a_i) z^i$.
2) Let X be as above a projective variety

$$X \subset \mathbb{P}^n_{\mathbb{C}}, X := \{z | f_i(z) = 0 \,\forall i\}.$$

The action of ϕ extends coordinatewise to $\mathbb{P}^n_{\mathbb{C}}$, and carries X to another variety, denoted X^ϕ, and called the **conjugate variety**. Since $f_i(z) = 0$ implies $\phi(f_i)(\phi(z)) = 0$, we see that

$$X^\phi = \{w | \phi(f_i)(w) = 0 \,\forall i\}.$$

If ϕ is complex conjugation, then it is clear that the variety X^ϕ that we obtain is diffeomorphic to X; but, in general, what happens when ϕ is not continuous?

Observe that, by the theorem of Steiniz, one has a surjection $\mathrm{Aut}(\mathbb{C}) \to \mathrm{Gal}(\bar{\mathbb{Q}}/\mathbb{Q})$, and by specialization the heart of the question concerns the action of $\mathrm{Gal}(\bar{\mathbb{Q}}/\mathbb{Q})$ on varieties X defined over $\bar{\mathbb{Q}}$.

For curves, since in general the dimensions of spaces of differential forms of a fixed degree and without poles are the same for X^ϕ and X, we shall obtain a curve of the same genus, hence X^ϕ and X are diffeomorphic.

6.1. Galois conjugates of projective classifying spaces

General questions of which the first is answered in the positive in most concrete cases, and the second is answered in the negative in many cases, as we shall see, are the following.

Question 6.2. Assume that X is a projective $K(\pi, 1)$, and assume $\phi \in \mathrm{Aut}(\mathbb{C})$.

A) Is then the conjugate variety X^ϕ still a classifying space $K(\pi', 1)$?

B) Is then $\pi_1(X^\phi) \cong \pi \cong \pi_1(X)$?

Since ϕ is never continuous, there would be no reason to expect a positive answer to both questions A) and B), except that Grothendieck showed ([58]).

Theorem 6.3. *Conjugate varieties X, X^ϕ have isomorphic algebraic fundamental groups*

$$\pi_1(X)^{alg} \cong \pi_1(X^\phi)^{alg},$$

where $\pi_1(X)^{alg}$ is the profinite completion of the topological fundamental group $\pi_1(X)$).

We recall once more that the profinite completion of a group G is the inverse limit

$$\hat{G} = \lim_{K \trianglelefteq_f G} (G/K),$$

of the factor groups G/K, K being a normal subgroup of finite index in G; and since finite index subgroups of the fundamental group correspond to finite unramified (étale) covers, Grothendieck defined in this way the algebraic fundamental group for varieties over other fields than the complex numbers, and also for more general schemes.

The main point of the proof of the above theorem is that if we have $f : Y \to X$ which is étale, also the Galois conjugate $f^\phi : Y^\phi \to X^\phi$ is étale (f^ϕ is just defined taking the Galois conjugate of the graph of f, a subvariety of $Y \times X$).

Since Galois conjugation gives an isomorphism of natural cohomology groups, which respects the cup product, as for instance the Dolbeault cohomology groups $H^p(\Omega^q_X)$, we obtain interesting consequences in the direction of question A) above. Recall the following definition.

Definition 6.4. Two varieties X, Y are said to be isogenous if there exist a third variety Z, and étale finite morphisms $f_X : Z \to X$, $f_Y : Z \to Y$.

Remark 6.5. It is obvious that if X is isogenous to Y, then X^ϕ is isogenous to Y^ϕ.

Theorem 6.6.

i) *If X is an Abelian variety, or isogenous to an Abelian variety, the same holds for any Galois conjugate X^ϕ.*
ii) *If S is a Kodaira fibred surface, then any Galois conjugate S^ϕ is also Kodaira fibred.*
iii) *If X is isogenous to a product of curves, the same holds for any Galois conjugate X^ϕ.*

Proof.

i) X is an Abelian variety if and only it is a projective variety and there is a morphism $X \times X \to X$, $(x, y) \mapsto (x \cdot y^{-1})$, which makes X a group (see [77], it follows indeed that the group is commutative). This property holds for X if and only if it holds for X^ϕ.

ii) The hypothesis is that there is $f : S \to B$ such that all the fibres are smooth and not all isomorphic: obviously the same property holds, after Galois conjugation, for $f^{\phi} : S^{\phi} \to B^{\phi}$.
iii) It suffices to show that the Galois conjugate of a product of curves is a product of curves. But since $X^{\phi} \times Y^{\phi} = (X \times Y)^{\phi}$ and the Galois conjugate of a curve C of genus g is again a curve of the same genus g, the statement follows. □

Proceeding with other projective $K(\pi, 1)$'s, the question becomes more subtle and we have to appeal to a famous theorem by Kazhdan on arithmetic varieties (see [41,42,64,65,70,91]).

Theorem 6.7. *Assume that X is a projective manifold with K_X ample, and that the universal covering $\tilde{X}$ is a bounded symmetric domain.*
Let $\tau \in \mathrm{Aut}(\mathbb{C})$ be an automorphism of $\mathbb{C}$.
Then the conjugate variety X^{τ} has universal covering $\tilde{X^{\tau}} \cong \tilde{X}$.

Simpler proofs follow from recent results obtained together with Antonio Di Scala, and based on the Aubin-Yau theorem. These results yield a precise characterization of varieties possessing a bounded symmetric domain as universal cover, and can be rather useful in view of the fact that our knowledge and classification of these fundamental groups is not so explicit.

We just mention the simplest result (see [41]).

Theorem 6.8. *Let X be a compact complex manifold of dimension n with K_X ample.*
Then the following two conditions (1) *and* (1′)*, resp.* (2) *and* (2′) *are equivalent:*

(1) *X admits a slope zero tensor $0 \neq \psi \in H^0(S^{mn}(\Omega^1_X)(-mK_X))$, (for some positive integer m);*
(1′) *$X \cong \Omega/\Gamma$, where Ω is a bounded symmetric domain of tube type and Γ is a cocompact discrete subgroup of $\mathrm{Aut}(\Omega)$ acting freely.*
(2) *X admits a semi special tensor $0 \neq \phi \in H^0(S^n(\Omega^1_X)(-K_X) \otimes \eta)$, where η is a 2-torsion invertible sheaf, such that there is a point $p \in X$ for which the corresponding hypersurface $F_p := \{\phi_p = 0\} \subset \mathbb{P}(TX_p)$ is reduced.*
(2′) *The universal cover of X is a polydisk.*

Moreover, in case (1)*, the degrees and the multiplicities of the irreducible factors of the polynomial ψ_p determine uniquely the universal covering $\widetilde{X} = \Omega$.*

6.2. Arithmetic of moduli spaces and faithful actions of the absolute Galois group

A basic remark is that all the schemes involved in the construction of the Gieseker moduli space are defined by equations involving only $\mathbb{Z}$-coefficients.

It follows that the absolute Galois group Gal$(\overline{\mathbb{Q}},\mathbb{Q})$ acts on the Gieseker moduli space $\mathfrak{M}_{a,b}$. In particular, it acts on the set of its irreducible components, and on the set of its connected components.

After an incomplete initial attempt in [8] in joint work with Ingrid Bauer and Fritz Grunewald, we were able in [9] to show:

Theorem 6.9. *The absolute Galois group* Gal$(\bar{\mathbb{Q}}/\mathbb{Q})$ *acts faithfully on the set of connected components of the Gieseker moduli space of surfaces of general type,*

$$\mathfrak{M} := \cup_{x,y \in \mathbb{N}, x,y \geq 1} \mathfrak{M}_{x,y}.$$

The main ingredients for the proof of theorem 6.9 are the following ones.

(1) Define, for any complex number $a \in \mathbb{C} \setminus \{-2g, 0, 1, \dots, 2g-1\}$, C_a as the hyperelliptic curve of genus $g \geq 3$ which is the smooth complete model of the affine curve of equation

$$w^2 = (z-a)(z+2g)\Pi_{i=0}^{2g-1}(z-i).$$

Consider then two complex numbers a, b such that $a \in \mathbb{C} \setminus \mathbb{Q}$: then $C_a \cong C_b$ if and only if $a = b$.

(2) If $a \in \bar{\mathbb{Q}}$, then by Belyi's theorem ([18]) there is a morphism $f_a : C_a \to \mathbb{P}^1$ which is branched only on three points, $0, 1, \infty$.

(3) The normal closure D_a of f_a yields a triangle curve, *i.e.*, a curve D_a with the action of a finite group G_a such that $D_a/G_a \cong \mathbb{P}^1$, and $D_a \to \mathbb{P}^1$ is branched only on three points.

(4) Take surfaces isogenous to a product $S = (D_a \times D')/G_a$ where the action of G_a on D' is free. Denote by $\mathcal{N}_a$ the union of connected components of the moduli space parametrizing such surfaces.

(5) Take all the possible twists of the G_a-action on $D_a \times D'$ via an automorphism $\psi \in \mathrm{Aut}(G_a)$ (*i.e.*, given the action $(x, y) \mapsto (\gamma x, \gamma y)$, consider all the actions of the form

$$(x, y) \mapsto (\gamma x, \psi(\gamma) y).$$

One observes that, for each ψ as above, we get more connected components in $\mathcal{N}_a$.

(6) Then an explicit calculation (using (4) and (5)) shows that the subgroup of $\mathrm{Gal}(\bar{\mathbb{Q}}/\mathbb{Q})$ acting trivially on the set of connected components of the moduli space would be a normal and abelian subgroup.
(7) Finally, this contradicts a known theorem (*cf.* [53]).

6.3. Change of fundamental group

Jean Pierre Serre proved in the 60's ([82]) the existence of a field automorphism $\phi \in \mathrm{Gal}(\bar{\mathbb{Q}}/\mathbb{Q})$, and a variety X defined over $\bar{\mathbb{Q}}$ such that X and the Galois conjugate variety X^ϕ have non isomorphic fundamental groups.

In [9] this phenomenon is vastly generalized, thus answering question B) in the negative.

Theorem 6.10. *If $\sigma \in \mathrm{Gal}(\bar{\mathbb{Q}}/\mathbb{Q})$ is not in the conjugacy class of complex conjugation, then there exists a surface isogenous to a product X such that X and the Galois conjugate surface X^σ have non-isomorphic fundamental groups.*

Since the argument for the above theorem is not constructive, let us observe that, in work in collaboration with Ingrid Bauer and Fritz Grunewald ([7,9]), we discovered wide classes of explicit algebraic surfaces isogenous to a product for which the same phenomenon holds.

References

[1] J. AMORÓS, MARC BURGER, A. CORLETTE, D. KOTSCHICK and D. TOLEDO, "Fundamental Groups of Compact Kähler Manifolds", Mathematical Surveys and Monographs, Vol. 44, Providence, RI: American Mathematical Society (AMS), 1966, xi, 140 pp.

[2] A. ANDREOTTI and T. FRANKEL, *The second Lefschetz theorem on hyperplane sections,* In: "Global Analysis (Papers in Honor of K. Kodaira)" Univ. Tokyo Press, Tokyo (1969), 1–20.

[3] T. AUBIN, *Équations du type Monge-Ampère sur les variétés kählériennes compactes*. Bull. Sci. Math. (2) **102** (1978), 63–95.

[4] G. BAGNERA and M. DE FRANCHIS, *Le superficie algebriche le quali ammettono una rappresentazione parametrica mediante funzioni iperellittiche di due argomenti*, Mem. di Mat. e di Fis. Soc. It. Sc. (3) **15** (1908), 253–343.

[5] W. BARTH, C. PETERS and A. VAN DE VEN, "Compact Complex Surfaces", Ergebnisse der Mathematik und ihrer Grenzgebiete (3), 4, Springer-Verlag, Berlin, 1984; second edition by W. Barth, K. Hulek, C. Peters, A. Van de Ven, Ergebnisse der Mathematik und ihrer Grenzgebiete, 3, Folge, A , 4, Springer-Verlag, Berlin, 2004.

[6] I. BAUER, F. CATANESE and F. GRUNEWALD, *Beauville surfaces without real structures*, In: "Geometric Methods in Algebra and Number Theory", Progr. Math., 235, Birkhäuser (2005), 1–42.

[7] I. BAUER, F. CATANESE and F. GRUNEWALD, *Chebycheff and Belyi polynomials, dessins d'enfants, Beauville surfaces and group theory*, Mediterranean J. Math. **3**, no. 2, (2006), 119–143.

[8] I. BAUER, F. CATANESE and F. GRUNEWALD, *The absolute Galois group acts faithfully on the connected components of the moduli space of surfaces of general type*, arXiv:0706.1466 ,13 pages.

[9] I. BAUER, F. CATANESE and F. GRUNEWALD, *Faithful actions of the absolute Galois group on connected components of moduli spaces* published online in Invent. Math. (2014) http://link.springer.com/article/10.1007/s00222-014-0531-2

[10] I. BAUER and F. CATANESE, *The moduli space of Keum-Naie surfaces*. Groups Geom. Dyn. **5** (2011), 231–250.

[11] I. BAUER and F. CATANESE, *Burniat surfaces I: fundamental groups and moduli of primary Burniat surfaces*, Classification of algebraic varieties, Schiermonnikoog, 2009. EMS Series of Congress Reports (2011), 49–76.

[12] I. BAUER and F. CATANESE, *Burniat surfaces. II. Secondary Burniat surfaces form three connected components of the moduli space*, Invent. Math. **180** (2010), no. 3, 559–588.

[13] I. BAUER and F. CATANESE, *Burniat surfaces III: deformations of automorphisms and extended Burniat surfaces*, Doc. Math. **18** (2013), 1089–1136.

[14] I. BAUER, F. CATANESE and R. PIGNATELLI, *Surfaces with geometric genus zero: a survey*, Ebeling, Wolfgang (ed.) *et al.*, Complex and differential geometry, Hannover, 2009, Springer, Proceedings in Mathematics 8 (2011), 1–48.

[15] I. BAUER and F. CATANESE, *Inoue type manifolds and Inoue surfaces: a connected component of the moduli space of surfaces with* $K^2 = 7$, $p_g = 0$, Geometry and arithmetic, EMS Ser. Congr. Rep., Eur. Math. Soc., Zürich (2012), 23–56.

[16] I. BAUER and F. CATANESE, *Burniat-type surfaces and a new family of surfaces with* $p_g = 0$, $K^2 = 3$ Rend. Circ. Mat. Palermo (2) **62** (2013), 37–60.

[17] I. BAUER, F. CATANESE and D. FRAPPORTI, *Generalized Burniat type surfaces*, arXiv:1409.1285.

[18] G. V. BELYĬ, *On Galois extensions of a maximal cyclotomic field*, Izv. Akad. Nauk SSSR Ser. Mat. 43:2 (1979), 269–276. Translation in Math. USSR- Izv. **14** (1980), 247–256.

[19] F. A. BOGOMOLOV, *The Brauer group of quotient spaces of linear representations*, Math. USSR-Izv. **30** (3) (1988), 455–485.
[20] A. BOREL, *Compact Clifford-Klein forms of symmetric spaces*, Topology **2** (1963), 111–122.
[21] Y. BRUNEBARBE, B. KLINGLER and B. TOTARO, *Symmetric differentials and the fundamental group*, Duke Math. J. **162** (2013), 2797–2813.
[22] J. A. CARLSON and D. TOLEDO, *Harmonic mappings of Kähler manifolds to locally symmetric spaces*, Inst. Hautes Études Sci. Publ. Math. No. **69** (1989), 173–201.
[23] E. CARTAN, "Lecons sur la géométrie des espaces de Riemann", Paris: Gauthier-Villars (Cahiers scientifiques publiés sous la direction de G. Julia, 2), VI, 1928, 273 pp.
[24] E. CARTAN, *Sur les domaines bornés homogénes de l'espace des n variables complexes*, Abhandl. Hamburg 11 (1935), 116–162.
[25] G. CASTELNUOVO, *Sulle superficie aventi il genere aritmetico negativo*, Palermo Rend. **20** (1905), 55–60.
[26] F. CATANESE, *On the Moduli Spaces of Surfaces of General Type*, J. Differential Geom. **19** (1984), 483–515.
[27] F. CATANESE, *Moduli and classification of irregular Kähler manifolds (and algebraic varieties) with Albanese general type fibrations*, Invent. Math. **104**, (1991), 263–289.
[28] F. CATANESE, *Compact complex manifolds bimeromorphic to tori*, In: "Abelian varieties" (Egloffstein, 1993), de Gruyter, Berlin (1995), 55–62.
[29] F. CATANESE, *Fibred surfaces, varieties isogenous to a product and related moduli spaces*, Amer. J. Math. **122**, no. 1 (2000), 1–44.
[30] F. CATANESE, *Moduli Spaces of Surfaces and Real Structures*, Ann. Math. (2) **158**, no. 2 (2003), 577–592.
[31] F. CATANESE, *Fibred Kähler and quasi projective groups*, Advances in Geometry, suppl., Special Issue dedicated to A. Barlotti's 80-th birthday (2003), Adv. Geom. suppl. (2003), S13–S27.
[32] F. CATANESE, *Deformation in the large of some complex manifolds, I*, Volume in Memory of Fabio Bardelli, Ann. Mat. Pura Appl. (4) **183** (2004), 261–289.
[33] F. CATANESE and B. WAJNRYB, *Diffeomorphism of simply connected algebraic surfaces*, J. Differential Geom. **76** (2007), no. 2, 177–213.
[34] F. CATANESE, *Differentiable and deformation type of algebraic surfaces, real and symplectic structures*, Symplectic 4-manifolds and algebraic surfaces, Lecture Notes in Math., 1938, Springer, Berlin, (2008), 55–167.

[35] F. CATANESE, K. OGUISO and T. PETERNELL, *On volume-preserving complex structures on real tori*, Kyoto J. Math. **50** (2010), no. 4, 753–775.

[36] F. CATANESE, *Irreducibility of the space of cyclic covers of algebraic curves of fixed numerical type and the irreducible components of* $Sing(\bar{\mathfrak{M}}_g)$, In: "Advances in Geometric Analysis", in honor of Shing-Tung Yau's 60th birthday, Advanced Lectures in Mathematics, Vol. 21, International Press and Higher Education Press (Beijing, China) (2012), 281–306.

[37] F. CATANESE, *A superficial working guide to deformations and moduli*, Handbook of moduli, Vol. I, Adv. Lect. Math. (ALM), 24, Int. Press, Somerville, MA, (2013), 161–215.

[38] F. CATANESE, *Topological methods in moduli theory*, arXiv:1411.3235.

[39] F. CATANESE, M. LÖNNE and F. PERRONI, *The irreducible components of the moduli space of dihedral covers of algebraic curves*, arXiv:1206.5498, Groups, Geometry and Dynamics, to appear.

[40] F. CATANESE, M. LÖNNE and F. PERRONI, *Genus stabilization for moduli of curves with symmetries*, arXiv:1301.4409.

[41] F. CATANESE and A. J. DI SCALA, *A characterization of varieties whose universal cover is the polydisk or a tube domain*, Math. Ann. **356** (2013), no. 2, 419–438.

[42] F. CATANESE and A. J. DI SCALA, *A characterization of varieties whose universal cover is a bounded symmetric domain without ball factors*, Adv. Math. **257** (2014), 567–580.

[43] M. T. CHAN and S. COUGHLAN, *Kulikov surfaces form a connected component of the moduli space*, Nagoya Math. Journal **210** (2013), 1–27.

[44] Y. CHEN, *A new family of surfaces of general type with* $K^2 = 7$ *and* $p_g = 0$, Math. Z. **275** (2013), 1275–1286.

[45] M. CORNALBA, *On the locus of curves with automorphisms*, Ann. Mat. Pura Appl. (4) **149** (1987), 135–151. Erratum: Ann. Mat. Pura Appl. (4) **187** (2008), 185–186.

[46] M. DE FRANCHIS, *Sulle superficie algebriche le quali contengono un fascio irrazionale di curve*, Palermo Rend. **20** (1905), 49–54.

[47] N. M. DUNFIELD and W. P. THURSTON, *Finite covers of random 3-manifolds*, Invent. Math. **166** (2006), 457–521.

[48] A. L. EDMONDS, *Surface symmetry. I*, Michigan Math. J. **29** (1982), 171–183.

[49] A. L. EDMONDS, *Surface symmetry. II*, Michigan Math. J. **30** (1983), 143–154.

[50] J. EELLS and J. H. SAMPSON, *Harmonic maps of Riemannian manifolds*, Amer. Jour. Math. **86** (1964), 109–160.

[51] F. ENRIQUES and F. SEVERI, *Mémoire sur les surfaces hyperelliptiques*, Acta Math. **32** (1909), 283–392 and **33** (1910), 321–403.

[52] P. EYSSIDIEUX, L. KATZARKOV, T. PANTEV and M. RAMACHANDRAN, *Linear Shafarevich conjecture*, Ann. of Math. (2) **176** (2012), 1545–1581.

[53] M. D. FRIED and M. JARDEN, "Field Arithmetic", Third edition, revised by Jarden, Ergebnisse der Mathematik und ihrer Grenzgebiete, 3, Folge, A Series of Modern Surveys in Mathematics, 11, Springer-Verlag, Berlin, 2008, xxiv+792 pp.

[54] M. D. FRIED and H. VÖLKLEIN, *The inverse Galois problem and rational points on moduli spaces*, Math. Ann. **290** (1991), 771–800.

[55] M. J. GREENBERG, "Lectures on Algebraic Topology", W. A. Benjamin, Inc., New York-Amsterdam, 1967, x+235 pp. Second edition: M. J. GREENBERG and J. R. HARPER, "Algebraic Topology. A First Course", Mathematics Lecture Note Series, 58. Benjamin/Cummings Publishing Co., Inc., Advanced Book Program, Reading, Mass., 1981, xi+311 pp.

[56] P. GRIFFITHS and J. MORGAN, "Rational Homotopy Theory and Differential Forms", Progress in Mathematics, Vol. 16, Birkhäuser, Boston, Mass., 1981, xi+242 pp.

[57] M. GROMOV, *Sur le groupe fondamental d'une variété kählérienne,* C. R. Acad. Sci., Paris, Sér. I **308** (1989), 67–70.

[58] A. GROTHENDIECK (dirigé par), "Revêtements étales et groupe fondamental", Séminaire de géométrie algébrique du Bois Marie 1960-61, Springer, Lecture Notes in Math., Vol. 224, 1971, Reedited by Société Mathématique de France in the series 'Documents mathématiques, 2003.

[59] J. L. HARER, *Stability of the homology of the mapping class groups of orientable surfaces*, Ann. of Math. (2) **121** (1985), 215–249.

[60] S. HELGASON, "Differential Geometry, Lie Groups, and Symmetric Spaces", Pure and Applied Mathematics, Vol. 80, Academic Press, Inc. [Harcourt Brace Jovanovich, Publishers], New York-London, 1978, xv+628 pp.

[61] H. HOPF, *Fundamentalgruppe und zweite Bettische Gruppe*, Comment. Math. Helv. **14** (1942), 257–309.

[62] J. H. HUBBARD, "Teichmüller theory and applications to geometry, topology, and dynamics. Vol. 1. Teichmüller theory", Matrix Editions, Ithaca, NY, 2006, xx+459 pp.

[63] M. INOUE, *Some new surfaces of general type*, Tokyo J. Math. **17** (1994), 295–319.

[64] D. KAZHDAN, *Arithmetic varieties and their fields of quasi-definition*, Actes du Congrès International des Mathematiciens (Nice, 1970), Tome 2, Gauthier-Villars, Paris (1971), 321 –325.

[65] D. KAZHDAN, *On arithmetic varieties. II*, Israel J. Math. **44** (1983), 139–159.

[66] S. P. KERCKHOFF, *The Nielsen realization problem*, Ann. of Math. (2) **117** (1983), 235–265.

[67] K. KODAIRA, *On Kähler varieties of restricted type (an intrinsic characterization of algebraic varieties)*, Ann. of Math. (2) **60** (1954), 28–48.

[68] H. LANGE, *Hyperelliptic varieties*, Tohoku Math. J. (2) **53** (2001), 491–510.

[69] C. LIVINGSTON, *Stabilizing surface symmetries*, Mich. Math. J. **32** (1985), 249–255.

[70] J. S. MILNE, *Kazhdan's Theorem on Arithmetic Varieties*, arXiv: math/0106197.

[71] J. W. MILNOR, "Morse Theory", Based on lecture notes by M. Spivak and R. Wells. Annals of Mathematics Studies, No. 51 Princeton University Press, Princeton, N.J., 1963, vi+153 pp.

[72] J. MILNOR, *Curvatures of left invariant metrics on Lie groups*, Adv. Math. **21** (3) (1976), 293–329.

[73] N. MOK, *The holomorphic or anti-holomorphic character of harmonic maps into irreducible compact quotients of polydiscs*, Math. Ann. **272** (1985), 197–216.

[74] N. MOK, *Strong rigidity of irreducible quotients of polydiscs of finite volume*, Math. Ann. **282** (1988), 555–578.

[75] J. MORGAN and G. TIAN, "Ricci Flow and the Poincaré Conjecture", Clay Mathematics Monographs 3. Providence, RI: American Mathematical Society (AMS); Cambridge, MA: Clay Mathematics Institute, 2007, xlii, 521 p.

[76] G. D. MOSTOW, "Strong Rigidity of Locally Symmetric Spaces", Annals of Mathematics Studies, no. 78, Princeton University Press, Princeton, N.J.; University of Tokyo Press, Tokyo, 1973.

[77] D. MUMFORD, "Abelian Varieties", Tata Institute of Fundamental Research Studies in Mathematics, No. 5 Published for the Tata Institute of Fundamental Research, Bombay; Oxford University Press, London, 1970, viii+242 pp.

[78] J. NIELSEN, *Untersuchungen zur Topologie der geschlossenen zweiseitigen Flächen I, II, III*, Acta Math. **50** (1927), 189–358, **53** (1929), 1–76, **58** (1932), 87–167.

[79] J. NIELSEN, *Abbildungsklassen endlicher Ordnung*, Acta Math. **75** (1943), 23–115.

[80] J. NIELSEN, *Surface transformation classes of algebraically finite type,* Danske Vid. Selsk. Math.-Phys. Medd. **21** (1944), 89 pp.
[81] J. H. SAMPSON, *Applications of harmonic maps to Kähler geometry*, Contemp. Math. **49** (1986), 125–134.
[82] J.-P. SERRE, *Exemples de variétés projectives conjuguées non homéomorphes*. C. R. Acad. Sci. Paris **258** (1964), 4194–4196.
[83] D. SINGERMAN, *Finitely maximal Fuchsian groups*, J. London Math. Soc. **6** (1972), 29–38.
[84] Y. T. SIU, *The complex-analyticity of harmonic maps and the strong rigidity of compact Kähler manifolds*, Ann. of Math. (2) **112** (1980), 73–111.
[85] Y. T. SIU, *Strong rigidity of compact quotients of exceptional bounded symmetric domains*, Duke Math. J. **48** (1981), 857–871.
[86] Y. T. SIU, *Strong rigidity for Kähler manifolds and the construction of bounded holomorphic functions*, Discrete groups in geometry and analysis, Pap. Hon. G. D. Mostow 60th Birthday, Prog. Math. **67** (1987), 124–151.
[87] E. H. SPANIER, "Algebraic Topology", McGraw-Hill Book Co., New York-Toronto, Ont.-London, 1966, xiv+528 pp.
[88] N. STEENROD, "The Topology of Fibre Bundles", Princeton Mathematical Series, vol. 14. Princeton University Press, Princeton, N. J., 1951, viii+224 pp.
[89] D. TOLEDO, *Projective varieties with non-residually finite fundamental group*, Inst. Hautes Études Sci. Publ. Math. No. **77** (1993), 103–119.
[90] A. TROMBA, *Dirichlet's energy on Teichmüller 's moduli space and the Nielsen realization problem*, Math. Z. **222** (1996), 451–464.
[91] VIEHWEG, E. AND ZUO, K.: *Arakelov inequalities and the uniformization of certain rigid Shimura varieties.*, J. Differential Geom. 77 (2007), 291– 352.
[92] C. VOISIN, *On the homotopy types of compact Kähler and complex projective manifolds,* Invent. Math. **157** (2004), 329–343.
[93] H. WEYL, *Invariants*, Duke Mathematical Journal **5** (3), (1939), 489–502.
[94] S. T. YAU, *Calabi's conjecture and some new results in algebraic geometry,* Proc. Natl. Acad. Sci. USA **74** (1977), 1798–1799.
[95] S. T. YAU, *A splitting theorem and an algebraic geometric characterization of locally Hermitian symmetric spaces*, Comm. Anal. Geom. **1** (1993), 473 – 486.
[96] B. ZIMMERMANN, *Surfaces and the second homology of a group*, Monatsh. Math. **104** (1987), 247–253.

Grothendieck at Pisa: crystals and Barsotti-Tate groups

Luc Illusie

1. Grothendieck at Pisa

Grothendieck visited Pisa twice, in 1966, and in 1969. It is on these occasions that he conceived his theory of crystalline cohomology and wrote foundations for the theory of deformations of p-divisible groups, which he called Barsotti-Tate groups. He did this in two letters, one to Tate, dated May 1966, and one to me, dated Dec. 2-4, 1969. Moreover, discussions with Barsotti that he had during his first visit led him to results and conjectures on specialization of Newton polygons, which he wrote in a letter to Barsotti, dated May 11, 1970.

May 1966 coincides with the end of the SGA 5 seminar [77]. Grothendieck was usually quite ahead of his seminars, thinking about questions which he might consider for future seminars, two or three years later. In this respect his correspondence with Serre [18] is fascinating. His local monodromy theorem, his theorems on good and semistable reduction of abelian varieties, his theory of vanishing cycles all appear in letters to Serre from 1964. This was to be the topic for SGA 7 [79], in 1967-68. The contents of SGA 6 [78] were for him basically old stuff (from before 1960), and I think that the year 1966-67 (the year of SGA 6) was a vacation of sorts for him, during which he let Berthelot and me quietly run (from the notes he had given to us and to the other contributors) a seminar which he must have considered as little more than an exercise.

In 1960 Dwork's proof [24] of the rationality of the zeta function of varieties over finite fields came as a surprise and drew attention to the power of p-adic analysis. In the early sixties, however, it was not p-adic analysis but étale cohomology which was in the limelight, due to its amazing development by Grothendieck and his collaborators in SGA 4 [76] and SGA 5 [77]. Étale cohomology provided a cohomological interpretation of the zeta function, and paved the way to a proof of the Weil conjectures. Moreover, it furnished interesting ℓ-adic Galois representations.

For example, if, say, X is proper and smooth over a number field k, with absolute Galois group $\Gamma_k = \mathrm{Gal}(\overline{k}/k)$, then for each prime number ℓ, the cohomology groups $H^i(X \otimes \overline{k}, \mathbf{Q}_\ell)$ are continuous, finite dimensional $\mathbf{Q}_\ell$-representations of Γ_k (of dimension b_i, the i-th Betti number of $X \otimes \mathbf{C}$, for any embedding $k \hookrightarrow \mathbf{C}$). These representations have local counterparts: for each finite place v of k and choice of an embedding of $\overline{k}$ in $\overline{k_v}$, the groups $H^i(X \otimes \overline{k_v}, \mathbf{Q}_\ell)$ are naturally identified to $H^i(X_{\overline{k}}, \mathbf{Q}_\ell)$, and the (continuous) action of the decomposition group $\Gamma_{k_v} = \mathrm{Gal}(\overline{k}_v/k_v) \subset \Gamma_k$ on them corresponds to the restriction of the action of Γ_k. For ℓ not dividing v, the structure of these local representations had been well known since 1964: by Grothendieck's local monodromy theorem, an open subgroup of the inertia group $I_v \subset \Gamma_{k_v}$ acts by unipotent automorphisms. For ℓ dividing v, the situation was much more complicated and it's only with the work of Fontaine in the 1970's and 1980's and the development of the so-called *p-adic Hodge theory* that a full understanding was reached. However, the first breakthroughs were made around 1965, with the pioneering work of Tate on p-divisible groups. Together with Dieudonné theory, this was one of the main sources of inspiration for Grothendieck's letters.

2. From formal groups to Barsotti-Tate groups

2.1. The Tate module of an abelian variety

As Serre explains in his Bourbaki talk [69], numerous properties of abelian varieties can be read from their group of division points. More precisely, if A is an abelian variety over a field k of characteristic p, $\overline{k}$ an algebraic closure of k, and ℓ a prime number, one can consider the *Tate module* of A,

$$T_\ell(A) := \varprojlim_n A(\overline{k})[\ell^n],$$

(where, for a positive integer m, $[m]$ denotes the kernel of the multiplication by m), which is a free module of rank r over $\mathbf{Z}_\ell$, equipped with a continuous action of $\mathrm{Gal}(\overline{k}/k)$. For $\ell \neq p$, one has $r = 2g$, where $g = \dim A$, and it has been known since Weil that when k is finite, this representation determines the zeta function of A. For k of characteristic $p > 0$, and $\ell = p$, one has $r \leq g$, and it was observed in the 1950's that in this case it was better to consider not just the kernels of the multiplications by p^n on the $\overline{k}$-points of A, but the *finite algebraic group schemes* $A[p^n]$, and especially their identity components $A[p^n]^0$, whose union is the *formal group* of A at the origin, a smooth commutative formal group of dimension g. For example, when $g = 1$ (A an elliptic curve), this group has dimension 1 and height 1 or 2 according to whether $r = 1$ (ordinary case) or $r = 0$ (supersingular case).

In the late 1950's and early 1960's formal groups were studied by Cartier, Dieudonné, Lazard, and Manin, mostly over perfect fields or sometimes over complete local noetherian rings with perfect residue fields. The notion of p-divisible group, which was first introduced by Barsotti [2] under the name "equidimensional hyperdomain", was formalized and studied by Serre and Tate (around 1963-66) before Grothendieck got interested in the topic. Let me briefly recall a few salient points of what was known at that time.

2.2. Dieudonné theory, p-divisible groups

Let k be a perfect field of characteristic $p > 0$, $W = W(k)$ the ring of Witt vectors on k, σ the automorphism of W defined by the absolute Frobenius of k, *i.e.* $a = (a_0, a_1, \cdots) \mapsto a^\sigma = (a_0^p, a_1^p, \cdots)$. Dieudonné theory associates with a finite commutative algebraic p-group G over k its *Dieudonné module*,

$$M(G) = \mathrm{Hom}(G, CW),$$

where CW is the fppf sheaf of Witt covectors on Spec k. This $M(G)$ is a W-module of finite length, equipped with a σ-linear operator F and a σ^{-1}-linear operator V satisfying the relation $FV = VF = p$, defined by the Frobenius F and the Verschiebung V on G. The above definition is due to Fontaine [31]. Classically (*cf.* [23,62]) one first defined $M(G)$ for G *unipotent* as $\mathrm{Hom}(G, \mathrm{CW}_u)$, where $\mathrm{CW}_u = \varinjlim W_n \subset \mathrm{CW}$ is the sheaf of unipotent covectors, and treated the multiplicative case by Cartier duality.

In general, by a *Dieudonné module*, one means a W-module, with operators F and V as above. The Dieudonné module of G is a contravariant functor of G, and this functor defines an anti-equivalence from the category of finite commutative algebraic p-groups over k to that of Dieudonné modules of finite length over W. The functor $G \mapsto M(G)$ is extended to formal groups, viewed as direct limits of connected finite commutative p-groups, and gives an embedding of the category of formal groups into a suitable category of Dieudonné modules.

A central result in the theory is the *Dieudonné-Manin classification theorem*, which describes the category of finitely generated Dieudonné modules up to isogeny. More precisely, let K denote the fraction field of W. Define an *F-space*[1] as a finite dimensional K-vector space equipped

[1] F-isocrystal, in today's terminology.

with a σ-linear automorphism F. The Dieudonné-Manin theorem says that, if k is algebraically closed, the category of F-spaces is semi-simple, and for each pair of integers (r, s), with $r = 0$ and $s = 1$ or $r \neq 0$ and $s > 0$ coprime, there is a unique isomorphism class of simple objects, represented by $E_{r,s} = K_\sigma[T]/(T^s - p^r)$, where $K_\sigma[T]$ is the non-commutative polynomial ring with the rule $Ta = a^\sigma T$, and F acts on $E_{r,s}$ by multiplication by T. Grothendieck was to revisit this theorem in his 1970 letter to Barsotti ([39], Appendix). We will discuss this in Section 5.

The Dieudonné module of a formal group is not necessarily finitely generated over W. For example, for the formal group $G = \widehat{\mathbf{G}_a}$, one has $M(G) = k_\sigma[[F]]$, with $V = 0$. Such phenomena do not occur, however, for p-divisible groups. Recall (cf. [69]) that given a base scheme S and an integer $h \geq 0$, a *p-divisible group* (*Barsotti-Tate group* in Grothendieck's terminology) *of height h* over S is a sequence of finite locally free commutative group schemes G_n of rank p^{nh} over S and homomorphisms $i_n : G_n \to G_{n+1}$, for $n \geq 1$, such that the sequences

$$0 \to G_n \xrightarrow{i_n} G_{n+1} \xrightarrow{p^n} G_{n+1}$$

are exact. For $n \geq 0$, $m \geq 0$, one then gets short exact sequences of group schemes over S

$$0 \to G_n \to G_{n+m} \xrightarrow{p^n} G_m \to 0.$$

The abelian sheaf $G := \varinjlim_n G_n$ on the fppf site S_{fppf} of S is then p-divisible, p-torsion, and $\operatorname{Ker} p.\mathrm{Id}_G$ is G_1, which in particular is finite locally free (of rank p^h). The sequence (G_n, i_n) is reconstructed from G by $G_n := \operatorname{Ker} p^n\mathrm{Id}_G$. It was to avoid confusion with the more general notion of p-divisible abelian sheaf - and also to pay tribute to Barsotti and Tate - that Grothendieck preferred the terminology *Barsotti-Tate group* to denote an abelian sheaf G on S_{fppf} which is p-divisible, p-torsion, and such that $\operatorname{Ker} p\mathrm{Id}_G$ is finite locally free.

The Cartier duals $G_n^\vee = Hom(G_n, \mathbf{G}_m)$, with the inclusions dual to $p : G_{n+1} \to G_n$, form a p-divisible group of height h, called the *dual* of G, denoted $G^\vee$. The basic examples of p-divisible groups are: $(\mathbf{Q}_p/\mathbf{Z}_p)_S = ((\mathbf{Z}/p^n\mathbf{Z})_S)$, its dual $(\mathbf{Q}_p/\mathbf{Z}_p)(1)_S = (\mu_{p^n,S})$, and the p-divisible group of an abelian scheme A over S

$$A[p^\infty] = (A[p^n])_{n\geq 1}.$$

When $S = \operatorname{Spec} K$, for K a field of characteristic $\neq p$, with an algebraic closure $\overline{K}$, a p-divisible group G of height h over S is determined by its

Tate module

$$T_p(G) := \varprojlim G_n(\overline{K}),$$

a free $\mathbf{Z}_p$-module of rank h, equipped with a continuous action of $\mathrm{Gal}(\overline{K}/K)$.

If $S = \mathrm{Spec}\, k$, with k as above, let G be a p-divisible group of height h over S. Then, G is determined by its *Dieudonné module*

$$M(G) := \varprojlim M(G_n),$$

a free W-module of rank h. And $M(G^\vee)$ is the dual $M(G)^\vee$, with F and V interchanged. The functor $G \mapsto M(G)$ is an (anti)-equivalence from the category of p-divisible groups over k to the full subcategory of Dieudonné modules consisting of modules which are free of finite rank over W.

Suppose now that $S = \mathrm{Spec}\, R$, where R is a complete discrete valuation ring, with perfect residue field k of characteristic $p > 0$ and fraction field K of characteristic zero, and let $\overline{K}$ be an algebraic closure of K. Let G be a p-divisible group of height h over S. Then two objects of quite a different nature are associated with G:

- the Tate module of G_K, $T_p(G_K)$, a free $\mathbf{Z}_p$-module of rank h on which $\mathrm{Gal}(\overline{K}/K)$ acts continuously
- the Dieudonné module of G_k, a free W-module of rank h, equipped with semi-linear operators F and V satisfying $FV = VF = p$.

Understanding the relations between these two objects, as well as their relations with the differential invariants associated with an abelian scheme A over S when $G = A[p^\infty]$ (such as $\mathrm{Lie}(A)$, its dual, and the de Rham cohomology group $H^1_{\mathrm{dR}}(A/S)$), was the starting point of p-adic Hodge theory.

2.3. The theorems of Tate and Serre-Tate

Let me briefly recall the main results, see [69] and [70] for details. Let $S = \mathrm{Spec}\, R, k, K, \overline{K}$ as before.

Theorem 2.3.1 (Tate) ([70, Theorem 4]).*The functor $G \mapsto G_K$ from the category of p-divisible groups over S to that of p-divisible groups over K is fully faithful,* i.e., *for p-divisible groups G, H over S, the map*

$$\mathrm{Hom}(G, H) \to \mathrm{Hom}(G_K, H_K)(\tilde{\to} \mathrm{Hom}_{\mathrm{Gal}(\overline{K}/K)})(T_p(G_K), T_p(H_K))$$

is bijective.

(Actually, Tate shows that 2.3.1 holds more generally for R local, complete, integral, normal, with perfect residue field k (of characteristic $p > 0$) and fraction field K of characteristic zero, but the proof is by reduction to the complete discrete valuation ring case.)

The equal characteristic analogue of 2.3.1 was to be established only many years later, by de Jong in 1998 [21].

Theorem 2.3.2 (Tate) ([70, Theorem 3, Corollary 2]). *Let $C := \widehat{\overline{K}}$ be the completion of $\overline{K}$, with its continuous action of $\mathrm{Gal}(\overline{K}/K)$. Let G be a p-divisible group over S. Then there is a natural decomposition, equivariant under $\mathrm{Gal}(\overline{K}/K)$,*

$$T_p(G_K) \otimes C \xrightarrow{\sim} (\mathrm{t}_G \otimes C(1)) \oplus ((\mathrm{t}_{G^\vee})^\vee \otimes C),$$

where, for a p-divisible group H over S, t_H denotes the Lie algebra of its identity component, a formal group over S.

Note that, in particular, if $d = \dim(\mathrm{t}_G)$, $d^\vee = \dim(\mathrm{t}_{G^\vee})$, one has $d + d^\vee = h$, where h is the height of G, a relation which can already be read on the Dieudonné modules of G_k and $G_k^\vee$.

When $G = A[p^\infty]$, for an abelian scheme A over S, the decomposition of 1.3.2 gives, by passing to the duals, a $\mathrm{Gal}(\overline{K}/K)$-equivariant decomposition of the form

$$H^1(A_{\overline{K}}, \mathbf{Z}_p) \otimes C \xrightarrow{\sim} (H^0(A_K, \Omega^1_{A_K/K}) \otimes C(-1)) \oplus (H^1(A_K, \mathcal{O}_{A_K}) \otimes C).$$

In his seminar at the Collège de France in 1966-67, Tate conjectured a generalization of this decomposition in higher dimension, the so-called *Hodge-Tate decomposition*, which was fully proven only in 1998, by Tsuji and de Jong as a corollary of the proof of Fontaine-Jannsen's conjecture C_{st} ([6,20,73]), after partial results by many authors (Raynaud, Fontaine, Bloch-Kato, Fontaine-Messing, Hyodo, Kato) (different proofs of C_{st} as well as of the related conjectures C_{cris}, C_{pst}, C_{dR} - by Faltings [27], Niziol [61], Beilinson [3,4] - have been given since then). A report on this is beyond the scope of these notes. In ([70, page 180]) Tate also asked for a similar decomposition for suitable rigid-analytic spaces over K. This question was recently solved by Scholze [67].

Theorem 2.3.3 (Serre-Tate). *Let R be a local artinian ring with perfect residue field k of characteristic $p > 0$, and let A_0 be an abelian variety over k. Then the functor associating with a lifting A of A_0 over R the corresponding lifting $A[p^\infty]$ of the p-divisible group $A_0[p^\infty]$ is an equivalence from the category of liftings of A_0 to that of liftings of $A_0[p^\infty]$.*

Serre and Tate did not write up their proof, sketched in notes of the Woods Hole summer school of 1964 [54]. The first written proof appeared in Messing's thesis ([58], V 2.3). A different proof was found by Drinfeld, see [48]. Another proof, based on Grothendieck's theory of deformations for Barsotti-Tate groups, is given in [43] (see 4.2 (ii)).

3. Grothendieck's letter to Tate: crystals and crystalline cohomology

That was roughly the state of the art when Grothendieck came on the scene. In the form of a riddle, a natural question was: what do the following objects have in common:

- a Dieudonné module
- a p-adic representation of the Galois group of a local field K as above
- a de Rham cohomology group?

At first sight, nothing. However, a p-adic Galois representation is, in a loose sense, some kind of analogue of a local system on a variety over $\mathbf{C}$. Local systems arising from the cohomology of proper smooth families can be interpreted in terms of relative de Rham cohomology groups, with their Gauss-Manin connection. In characteristic zero at least, integrable connections correspond to compatible systems of isomorphisms between stalks at infinitesimally near points. On the other hand, by Oda's thesis [62] the reduction mod p of the Dieudonné module of the p-divisible group $A[p^\infty]$ of an abelian variety A over k is isomorphic to $H^1_{\mathrm{dR}}(A/k)$ (see the end of this section). Recall also that in his letter to Atiyah (Oct. 14, 1963) [34] Grothendieck had asked for an algebraic interpretation of the Gauss-Manin connection in the proper smooth case and discussed the H^1_{dR} of abelian schemes. These were probably some of the ideas that Grothendieck had in mind when he wrote his famous letter to Tate, of May 1966. Here is the beginning of this letter:

"Cher John,

J'ai réfléchi aux groupes formels et à la cohomologie de de Rham, et suis arrivé à un projet de théorie, ou plutôt de début de théorie, que j'ai envie de t'exposer, pour me clarifier les idées.

Chapitre 1 La notion de cristal.
Commentaire terminologique : Un cristal possède deux propriétés caractéristiques : la *rigidité*, et la faculté de *croître* dans un voisinage approprié. Il y a des cristaux de toute espèce de substance : des cristaux de soude, de soufre, de modules, d'anneaux, de schémas relatifs etc."

Grothendieck refined and expanded his letter in a seminar he gave at the I.H.É.S. in December, 1966, whose notes were written up by Coates and Jussila [35]. The contents are roughly the following:

3.1. Crystals

The word is as beautiful as the mathematical objects themselves. Starting with a scheme S over a base T, Grothendieck considers the category $\mathcal{C}$ of T-thickenings of open subschemes of S, *i.e.* T-morphisms $i : U \hookrightarrow V$ where U is an open subscheme of S and i is a locally nilpotent closed immersion, with maps from $U \hookrightarrow V$ to $U' \hookrightarrow V'$ given by the obvious commutative diagrams. He calls *crystal in modules* on S (relative to T) a cartesian section over $\mathcal{C}$ of the fibered category of quasi-coherent modules over the category $\mathrm{Sch}_{/T}$ of T-schemes. More generally, given a fibered category $\mathcal{F}$ over $\mathrm{Sch}_{/T}$, he calls crystal in objects of $\mathcal{F}$ a cartesian section of $\mathcal{F}$ over $\mathcal{C}$. He gives a few examples (especially for T of characteristic $p > 0$, showing that crystals in modules with additional Frobenius and Verschiebung structures can be viewed as families of Dieudonné modules parametrized by the points of S) and, for S smooth over T, he gives a description of a crystal in modules in terms of a quasi-coherent module E on S equipped with what we now call a *stratification*, namely an isomorphism $\chi : p_1^*E \xrightarrow{\sim} p_2^*E$, where p_1, p_2 are the projections to X from the formal completion of the diagonal $S \hookrightarrow S \times_T S$, such χ satisfying a natural cocycle descent condition on the formal completion of the diagonal in $S \times_T S \times_T S$.

He also introduces the first avatar of what was to become the *crystalline site*, which he calls the "crystallogenic site" ("site cristallogène"), consisting of thickenings $U \hookrightarrow V$ as above, with covering families those families $(U_i \hookrightarrow V_i)_{i \in I} \to (U \hookrightarrow V$ such that $(V_i \to V)$ is Zariski covering and $U_i = U \cap V_i$. He notes that crystals in modules can be re-interpreted in terms of certain sheaves on this site, and that for $f : X \to Y$ (in $\mathrm{Sch}_{/T}$), the functoriality will not be for the sites, but for the corresponding topoi (a point which will later be crucial in Berthelot's theory [5]). He adds that he expects that for f proper and smooth, then the $R^i f_*$ of the structural sheaf of the crystallogenic topos of X should give the relative de Rham cohomology sheaves $R^i f_* \Omega^\cdot_{X/Y}$ endowed with their Gauss-Manin connection.

However, in the course of his letter Grothendieck realizes that these definitions will have to be modified to take into account characteristic $p > 0$ phenomena. He adds a handwritten note in the margin: "fait mouche en car. 0, et pas en car. $p > 0$". We will discuss this in the next section.

3.2. De Rham cohomology as a crystal

This is the most striking observation. In his letter, Grothendieck, carefully enough, writes: “Chapitre 2 : la cohomologie de de Rham est un cristal. L’affirmation du titre n’est pour l’instant qu’une hypothèse ou un vœu pieux, mais je suis convaincu qu’elle est essentiellement correcte.” He gives two pieces of evidence for his claim.

(a) He mentions Monsky-Wahnitzer’s work on the independence of de Rham cohomology of (“weakly complete”) liftings to $W(k)$ of smooth affine k-schemes. He criticizes the authors for not being able to globalize their construction to proper schemes (except for curves) and having to work $\otimes\mathbf{Q}$. A couple of years later, Berthelot’s thesis solved the globalization problem. However a full understanding of Monsky-Washnitzer cohomology was only reached in the 1980’s, by Berthelot again, with his theory of *rigid cohomology* (where $\otimes\mathbf{Q}$ is essential).

(b) He says that he has found an algebraic construction of the Gauss-Manin connection on $R^i f_*\Omega^\bullet_{X/S}$ for a smooth morphism X/S (and S over some base T), or rather on the object $Rf_*\Omega^\bullet_{X/S}$ of the derived category $D(S, \mathcal{O}_S)$, adding that, however, he has not yet checked the integrability condition. He also asks for a crystalline interpretation (*i.e.* in terms of cohomology of a suitable crystalline site) of this connection, and of the corresponding Leray spectral sequence for $X \to S \to T$. In his lectures at the I.H.É.S. [35], he gave the details of his construction and explained the link between (a) and (b). His construction, close in spirit to Manin’s, based on local liftings of derivations, was used later by Katz in [47]. As for the integrability, Katz and Oda found a simple, direct proof in [46], based on the analysis of the Koszul filtration of the absolute de Rham complex $\Omega^\bullet_{X/T}$. However, the crystalline interpretation requested by Grothendieck, which was to be given by Berthelot in his thesis [5], and the (dual) approach, in characteristic zero, via $\mathcal{D}$-modules was to give a deeper insight into this structure.

As for the link between (a) and (b), Grothendieck’s observation was the following. Suppose X is proper and smooth over $S = \operatorname{Spec} W[[t]]$ ($t = (t_1, \cdots, t_n)$), and $\mathcal{H}^i_{dR}(X/S)$ is free of finite type for all i. Let $u : \operatorname{Spec} W \to S$, $v : \operatorname{Spec} W \to S$ be sections of S such that $u \equiv v \bmod p$. We then get two schemes over W, $X_u := u^*X$, $X_v := v^*X$ such that $X_u \otimes k = X_v \otimes k = Y$, and two de Rham cohomology groups, $H^i_{dR}(X_u/W) = u^*\mathcal{H}^i(X/S)$, $H^i_{dR}(X_v/W) = v^*\mathcal{H}^i(X/S)$. By the Gauss-Manin connection

$$\nabla : \mathcal{H}^i_{dR}(X/S) \to \Omega^1_{S/W} \otimes \mathcal{H}^i_{dR}(X/S),$$

we get an *isomorphism*

$$\chi(u,v) : H^i_{dR}(X_u/W) \xrightarrow{\sim} H^i_{dR}(X_v/W),$$

defined by

$$u^*(a) \mapsto \sum_{m\geq 0}(1/m!)(u^*(t) - v^*(t))^m v^*(\nabla(D)^m a)$$

for $a \in \mathcal{H}^i_{dR}(X/S)$, with the usual contracted notations, where

$$D = (D_1, \cdots, D_n),\ \ D_i = \partial/\partial t_i$$

(note that $(1/m!)(u^*(t) - v^*(t))^m \in W$ and that the series converges p-adically: this is easy for $p > 2$, was proved by Berthelot in general [5]). These isomorphisms satisfy $\chi(u,u) = \mathrm{Id}$ and $\chi(v,w)\chi(u,v) = \chi(u,w)$, for $w \equiv u \bmod p$.

This suggested to Grothendieck that, for Y/k proper, smooth, given two proper smooth liftings X_1, X_2 of Y over W, one could hope for an isomorphism (generalizing $\chi(u,v)$)

$$\chi_{12} : H^i_{dR}(X_1/W) \xrightarrow{\sim} H^i_{dR}(X_2/W)$$

with $\chi_{23}\chi_{12} = \chi_{13}$. Monsky-Washnitzer's theory provided such an isomorphism (after tensoring with $\mathbf{Q}$) in the affine case, for good liftings X_i. This hope was to be realized by the construction of *crystalline cohomology* groups $H^i(Y/W)$ (depending only on Y, with no assumption of existence of lifting), providing a canonical isomorphism:

$$\chi : H^i(Y/W) \xrightarrow{\sim} H^i_{dR}(X/W)$$

for any proper smooth lifting X/W of Y, such that for X_1, X_2 as above, $\chi_2 = \chi_{12}\chi_1$. Grothendieck sketched the construction in [35] (which worked for $p > 2$), the general case was done and treated in detail by Berthelot [5].

But let us come back to Grothendieck's letter. For $f : X \to S$ proper and smooth, of relative dimension d, S being over some base T, Grothendieck (boldly) conjectures:

(*) $Rf_*\Omega^\bullet_{X/S}$ should be a perfect complex on S, underlying a structure of crystal relative to T, commuting with base change, and for each prime number p, on the corresponding object H_p for the reduction mod p of f, the Frobenius operator $F : H_p^{(p)} \to H_p$ should be an *isogeny*, with an operator V in the other direction, satisfying $FV = VF = p^d$.

He analyzes the case where X/S is an abelian scheme, and makes two critical observations.

(3.2.1) For $S = \text{Spec } W$ and X/S an abelian scheme, $H^1_{\text{dR}}(X/S)$, which should be the value on S of the sought after *crystal* defined by $X \otimes k$, equipped with the operators F and V defined by Frobenius and Verschiebung, should be the *Dieudonné module* of the p-divisible group associated with $X \otimes k$.

(3.2.2) He realizes that in char. $p > 0$ his assertion that de Rham cohomology is a crystal in the sense defined at the beginning of his letter is wrong. In fact, for S smooth over T, a (quasi-coherent) crystal on S/T would correspond (cf. 3.1) to a *stratified module* M relative to T. And it would not be possible to put such a stratification on $\mathcal{H}^1_{\text{dR}}(A/S)$ relative to $T = \text{Spec } k$ for an *elliptic curve* A over S, in a way which would be functorial in A and compatible with base change, for S of finite type, regular, and of dimension ≤ 1 over k. He gives the example of an elliptic curve A/S, S a smooth curve over T, with a rational point s where A_s has Hasse invariant zero. In this case, the (absolute) Frobenius map $F : H^1(A_s, \mathcal{O})^{(p)} \to H^1(A_s, \mathcal{O})$ is zero, so $F^2 : H^1_{\text{dR}}(A_s/s)^{(p^2)} \to H^1_{\text{dR}}(A_s/s)$ is zero, hence, because of the stratification, F^2 would be zero on the completion of S at s, hence in a neighborhood of s, which is not the case when S is modular. This observation led him, in [35], to call the previously defined site (the "crystallogenic site") the *infinitesimal site*, and define a new site (and, accordingly, a new notion of crystal), putting *divided powers* on the ideals of the thickenings. Technical problems arose for $p = 2$, as the natural divided powers of the ideal pW are not p-adically nilpotent, but these were later solved by Berthelot in his thesis, dropping the restriction of nilpotence on the divided powers introduced in [35], on schemes where p is locally nilpotent.

Why add divided powers? In [35] Grothendieck explains that the introduction of divided powers was "practically imposed by the need to define the first Chern class $c_1(L) \in H^2(X_{\text{cris}}, \mathcal{J})$ of an invertible sheaf L on X", as the obstruction to lifting L to X_{cris}, using the logarithm

$$\log : 1 + \mathcal{J} \to \mathcal{J}, \ \log(1+x) = \sum_{n \geq 1} (-1)^{n-1}(n-1)!(x^n/n!),$$

where $\mathcal{J}$ is the kernel of the natural surjective map from $\mathcal{O}_{X_{\text{cris}}}$ to $\mathcal{O}_{X_{\text{zar}}}$. While this was certainly a motivation, it seems to me that the primary motivation was to make de Rham cohomology a crystal.

For S smooth over T, and E a quasi-coherent module on S, a stratification on E relative to T is given by an action of the ring $\text{Diff}_{S/T} =$

$\cup \mathrm{Diff}^n_{S/T}$ of differential operators of S over T. In general, an integrable connection ∇ on E relative to T does not extend to an action of $\mathrm{Diff}_{S/T}$. But it does extend to an action of the ring of *PD-differential operators* $\text{PD-Diff}_{S/T} = \cup\, \text{PD-Diff}^n_{S/T}$ (PD for "puissances divisées"), where PD-$\mathrm{Diff}^n_{S/T}$ is the dual (with values in $\mathcal{O}_S$) of the *divided power envelope* $D^n_{S/T}$ of the ideal of the n-th infinitesimal neighborhood of S in $S \times_T S$. In terms of local coordinates $(x_i)_{1 \leq i \leq r}$ on S, the associated graded (for the filtration by the order) of Diff is a divided power polynomial algebra on generators δ_i corresponding to $\partial/\partial x_i$, while that of PD-Diff is a usual polynomial algebra on the δ_i's. And crystals, for a suitable site defined by thickenings with divided powers, were to correspond exactly (for S/T smooth and schemes T where p is locally nilpotent) to modules with an integrable connection (satisfying an additional condition of *p-nilpotency*), the datum of ∇ being equivalent to that of the action of $\text{PD-Diff}_{S/T}$, *i.e.* to the PD-analogue of a stratification relative to T.

A good definition of a *crystalline site* was worked out by Berthelot in [5], and the first part of Grothendieck's conjecture (*) above proven in ([5], V 3.6). The existence of V satisfying $FV = VF = p^d$ was shown by Berthelot-Ogus in ([11], 1.6).

As for Grothendieck's expectation (3.2.1) above, it was proved by Oda ([62], 5.11) that $H^1_{\mathrm{dR}}(A/k)(\xrightarrow{\sim} H^1_{\mathrm{dR}}(X/S) \otimes k)$ (where $A := X \otimes k$) is isomorphic to the Dieudonné module of $A[p]$, *i.e.* to M/pM, where M is the Dieudonné module of the p-divisible group G associated with A. With Berthelot's definition of crystalline cohomology, $H^1_{\mathrm{dR}}(X/S)$ is isomorphic to the crystalline cohomology group $H^1(A/W)$. The isomorphism between $H^1(A/W)$ (with the operators F and V) and M was first proved by Mazur-Messing in [56]. Different proofs were given later ([8], ([42], II 3.11.2)).

For a survey of crystalline cohomology (up to 1990), see [44].

4. Grothendieck's letter to Illusie: deformations of Barsotti-Tate groups

This letter has two parts. In the first part, Grothendieck describes a (conjectural) theory of first order deformations for flat commutative group schemes. This theory was developed in the second volume of my thesis [41]. In the second part, he applies it to Barsotti-Tate groups, stating theorems of existence and classification of deformations for Barsotti-Tate groups and truncated ones. He gave the proofs in his course at the Collège de France in 1970-71. These proofs were written up in [43].

4.1. Deformations of flat commutative group schemes.

"Marina, les 2-4 déc. 1969. Cher Illusie, Le travail avance, mais avec une lenteur ridicule. J'en suis encore aux préliminaires sur les groupes de Barsotti-Tate sur une base quelconque - il n'est pas encore question de mettre des puissances divisées dans le coup ! La raison de cette lenteur réside en partie dans le manque de fondements divers. (...) De plus, à certains moments, je suis obligé d'utiliser une théorie de déformations pour des schémas en groupes plats mais non lisses, qui doit certainement être correcte, et qui devrait sans doute figurer dans ta thèse, mais que tu n'as pas dû écrire encore, sans doute. Je vais donc commencer par te soumettre ce que tu devrais bien prouver. (...)"

Grothendieck then proposes a theory of (first-order) deformations for group schemes G/S which are locally of finite presentation and flat. He says he is mainly interested in the commutative case, but that the non commutative case should also be considered (which I did in [41]). In both cases, the invariant which controls the deformations is the *co-Lie complex* of G/S,

$$\ell_G := Le^* L_{G/S},$$

where $e : S \to G$ is the unit section, $L_{G/S}$ the cotangent complex of G/S, and $Le^* : D^-(G, \mathcal{O}_G) \to D^-(S, \mathcal{O}_S)$ the derived functor of e^*. This complex appeared for the first time in the work of Mazur-Roberts [57]. As G/S is locally a complete intersection, this is a perfect complex, of perfect amplitude in $[-1, 0]$ (and $L_{G/S}$ is recovered from it by $L_{G/S} = \pi^* \ell_G$, where $\pi : G \to S$ is the projection). When G is *commutative, finite and locally free*, ℓ_G is related to the Cartier dual G^* of G by the following beautiful formula (proposed by Grothendieck in his letter, and proven by him in his course at Collège de France, see ([56, 14.1])): if M is a quasi-coherent $\mathcal{O}_S$-module, there is a natural isomorphism (in $D(S, \mathcal{O}_S)$),

$$R\mathcal{H}om_{\mathcal{O}_S}(\ell_G, M) \xrightarrow{\sim} \tau_{\leq 1} R\mathcal{H}om(G^*, M).$$

After having stated a (conjectural) theory of obstructions for deformations of G in the commutative case, Grothendieck realizes that he needs more. In fact, he sees that he will need to apply this theory to the truncations $G(n) = \operatorname{Ker} p^n \mathrm{Id}_G$ of Barsotti-Tate groups G. Such truncations are $\mathbf{Z}/p^n$-modules, and deformations should preserve this structure. But then, in order to get a common theory for commutative group schemes and commutative group schemes annihilated by p^n, it is natural to introduce a ring A of *complex multiplication* acting on G (hence on ℓ_G);

obstruction groups (and related ones) should involve this A-linear structure. He adds that whether A should be a constant ring or a more or less arbitrary sheaf of rings, he has not yet tried to think about it. In [41] I treat the case where A is a (non necessarily commutative) ring scheme satisfying a mild hypothesis with respect to the given infinitesimal thickening. For applications to deformations of abelian schemes and Barsotti-Tate groups, the case where A is the scheme defined by a constant commutative ring (in fact, $\mathbf{Z}$ or $\mathbf{Z}/n$) suffices. One of the main results is the following.

Theorem 4.1.1 ([41], VII 4.2.1). *Let $i : S \hookrightarrow S'$ be a closed immersion defined by an ideal I of square zero. Let A be a (constant) ring, and G a scheme in A-modules over S, flat and locally of finite presentation over S. Let us work with the fppf topology on S. Consider the differential graded ring $A \otimes^L_{\mathbf{Z}} \mathcal{O}_S$* [2] *and let*

$$\ell^{\vee}_G \in D^{[0,1]}(A \otimes^L_{\mathbf{Z}} \mathcal{O}_S)$$

be the Lie complex of G/S, defined in ([41, VII (4.1.5.4)]), *whose image in $D(\mathcal{O}_S)$ is the dual $R\mathcal{H}om(\ell_G, \mathcal{O}_S)$ of the co-Lie complex of G. Then:*

(i) *There is an obstruction*

$$o(G, i) \in \mathrm{Ext}^2_A(G, \ell^{\vee}_G \otimes^L_{\mathcal{O}_S} I)$$

whose vanishing is necessary and sufficient for the existence of a deformation G' of G over S' as a scheme in A-modules, flat over S'.

(ii) *This obstruction depends functorially on G in the following sense: if $u : F \to G$ is a homomorphism of (flat and locally of finite presentation) schemes in A-modules over S, then*

$$u^*(o(G, i)) = \ell^{\vee}_u(o(F, i)),$$

where u^* and $\ell^{\vee}_u$ are the natural functoriality maps.

(iii) *When $o(G, i) = 0$, the set of isomorphism classes of deformations G' of G over S' is a torsor under $\mathrm{Ext}^1_A(G, \ell^{\vee}_G \otimes^L_{\mathcal{O}_S} I)$, and the group of automorphisms of a given deformation G' is $\mathrm{Ext}^0_A(G, \ell^{\vee}_G \otimes^L_{\mathcal{O}_S} I)$.*

The proof is long and technical, involving complicated diagrams and a delicate homotopical stabilization process. Those diagrams were suggested by Grothendieck's attempts to calculate $\mathrm{Ext}^i_{\mathbf{Z}}(G, -)$ by *canonical*

[2] When $A = \mathbf{Z}/n$, this is simply the differential graded ring $[\mathcal{O}_S \xrightarrow{n} \mathcal{O}_S]$, in degrees 0 and -1.

resolutions of G by finite sums of $\mathbf{Z}[G^r]$ ($r \geq 1$), which he had described, for $i \leq 2$, in ([37, VII 3.5]), and that he recalls in his letter to me (such resolutions, called *Moore complexes*, were constructed by Deligne [22]). Variants, called *MacLane resolutions*, involving sums of $\mathbf{Z}[G^r \times \mathbf{Z}^s]$, and taking into account the multiplicative structures, are given in ([41, VI 11.4.4])) (see also [13]). The method, however, is flexible, and can be applied to many other kinds of deformation problems (such as morphisms, with source and target, or only source or target extended). The functoriality statement (iii) (and similar properties for homomorphisms of rings $A \to B$) are of course crucial in the applications, where the obstruction group Ext^2 may be nonzero, but the obstruction is zero, because of functoriality constraints.

At the end of ([41, VII]) I write that Deligne's theory of Picard stacks might yield simpler proofs of the above results. But over forty years have elapsed, and no such simpler proof has yet appeared.

4.2. Deformations of BT's and BT_n's

The second part of his letter starts with what Grothendieck calls "fascicule de résultats sur les groupes de BT (= Barsotti-Tate) et les groupes de BT tronqués ("part soritale")". As he had explained in the first part, while the goal was to show the existence of infinitesimal liftings of BT's and classify them, the key objects of study were in fact n-truncated BT's.

Given an integer $n \geq 1$, an *n-truncated BT* (or BT_n) G over a base scheme S is defined as an abelian sheaf on S (for the fppf topology), which is annihilated by p^n, flat over $\mathbf{Z}/p^n$, and such that $G(1) := \mathrm{Ker}\, p\mathrm{Id}_G$ is finite locally free over S (its rank is then of the form p^h, where h is called the height of G). When $n = 1$, one imposes an additional condition, namely that on $S_0 = V(p) \subset S$, one has $\mathrm{Ker}\, V = \mathrm{Im}\, F$, where $V : G_0^{(p)} \to G_0$ and $F : G_0 \to G_0^{(p)}$ are the Verschiebung and the Frobenius morphisms respectively, with $G_0 = G \times_S S_0$.

If G is a BT over S, one shows that for all $n \geq 1$, $G(n) := \mathrm{Ker}\, p^n\mathrm{Id}_G$ is a BT_n, which raises the question whether any BT_n is of the form $G(n)$ for a BT G. Grothendieck tackles this question in his letter, simultaneously with that of existence and classification of deformations of BT's and BT_n's. In order to apply the general obstruction theory, one needs precise information on co-Lie complexes of truncated BT's. If S is a scheme where $p^N\mathcal{O}_S = 0$, for an integer $N \geq 1$, and G is a BT_n, with $n \geq N$, then the co-Lie complex ℓ_G enjoys nice properties. In particular $\omega_G := \mathcal{H}^0(\ell_G)$ and $n_G := \mathcal{H}^{-1}(\ell_G)$ are locally free of the same rank, which, when $G = \mathcal{G}(n)$ for a BT $\mathcal{G}$ on S, is the dimension of the formal Lie group associated to $\mathcal{G}$. After a subtle analysis of the relations between these invariants and the behavior of $\text{Ext}^*(-, M)$, M a quasi-coherent $\mathcal{O}_S$-module, under exact sequences $0 \to G(n) \to G(n+m) \to G(m) \to 0$, Grothendieck obtains the following theorem (first stated in ([36, 6.3])), see ([39, VI 4.1]), ([43, 4.4])):

Theorem 4.2.1. *Let $n \geq 1$. Let $i : S \to S'$ be a nilimmersion, with S' affine.*

(1) *Let G be a* BT_n *on S. There exists a* BT_n *G' on S' extending G.*
(2) *Let H be a BT on S. There exists a BT H' on S' extending H.*
(3) *If $E(H, S')$ (resp. $E(H(n), S')$) denotes the set of isomorphism classes of BT's (resp.* BT_n*'s) on S' extending H (resp. $H(n)$)), then the natural map*

$$E(H, S') \to E(H(n), S')$$

is surjective, and bijective if i is nilpotent of level k and there exists $N \geq 1$ such that $p^N \mathcal{O}_S = 0$ and $n \geq kN$.
(4) *For $k = 1$ and $n \geq N$ as in* (3)*, $E(G, S')$ (resp. $E(H, S')$) is a torsor under $t_{G^\vee} \otimes t_G \otimes I$ (resp. $t_{H^\vee} \otimes t_H \otimes I$), where I is the ideal of i, and the automorphism group of a deformation of G (resp. H) on S' is $t_{G^\vee} \otimes t_G \otimes I$ (resp. 0). Here t_G denotes the dual of ω_G, and $G^\vee$ the Cartier dual of G.*
(5) *If S is the spectrum of a complete noetherian local ring with perfect residue field, for any* BT_n *G on S there exists a BT H on S such that $G = H(n)$.*

(For *formal* BT's H, (2) and (4) had been proven by Cartier and Lazard [17,53].)

Just to give an idea on how 4.2.1 is derived from 4.1.1, let me observe that assertions (4) for a BT H follow from the following facts:

- deforming H on S' is equivalent to deforming the projective system $H(\cdot) = H(n)_{n \geq 1}$ on S',
- the obstruction $o(H, i)$ to deforming $H(.)$ lies in

$$\text{Ext}^2_{\mathbf{Z}/p^\cdot}(H(\cdot), \ell^\vee_{H(\cdot)} \otimes^L_{\mathcal{O}_S} I),$$

  and this group is zero,
- the isomorphism classes of deformations of H on S' form a torsor under

$$\text{Ext}^1_{\mathbf{Z}/p^\cdot}(H(\cdot), \ell^\vee_{H(\cdot)} \otimes^L_{\mathcal{O}_S} I),$$

  and this group is canonically isomorphic to $t_{H^\vee} \otimes t_H \otimes I$,

- the automorphism group of a deformation H' on S' is

$$\mathrm{Ext}^0_{\mathbf{Z}/p^\cdot}(H(\cdot), \ell^\vee_{H(\cdot)} \otimes^L_{\mathcal{O}_S} I),$$

which is zero.

This theorem had immediate applications, and cast long shadows.

Among the immediate applications, we have (i) and (ii) below, already mentioned by Grothendieck in his letter:

(i) The *pro-representability of the deformation functor* of a BT H over a perfect field k of characteristic $p > 0$, namely the fact that the functor of deformations of H over the category of artinian local $W(k)$-algebras of residue field k is pro-represented by a smooth formal scheme

$$S = \mathrm{Spf}\,(W(k)[[t_{ij}]]_{1 \leq i \leq d, 1 \leq j \leq d^\vee})$$

where d ($= \mathrm{rk}\,\omega_H$) is the dimension of H, and $d^\vee$ that of its Cartier dual $H^\vee$ (as after 2.3.2, one has $d + d^\vee = h$, where h is the height of H).

(ii) A *short proof of the existence of infinitesimal liftings of abelian schemes and of Serre-Tate's theorem 1.3.3*, see ([43], Appendice). Concerning Serre-Tate's theorem, Grothendieck made an interesting comment. At the beginning of [48], Katz recalls that this theorem, in the case of a g-dimensional *ordinary* abelian variety over k (assumed to be algebraically closed), implies the existence of "a remarkable and unexpected structure of *group* on the corresponding formal moduli space". At the end of his letter (6.7), Grothendieck explains why, in fact, this structure was expected. His explanation relies on a theory of deformations of *extensions* in the general context of flat group schemes in A-modules (as in 4.1.1), which he applies to BT's or BT_n's. Unfortunately, this (beautiful) part of his letter was not discussed in [41] nor [43].

As for the long shadows:

(iii) Property 4.2.1 (5) (which did not appear in [36] nor [39], but was an easy consequence of the theory) implies a *formula for the different* $d_{G/S}$ of a BT_n G over the spectrum S of a complete discrete valuation ring R with perfect residue field k of characteristic $p > 0$, with dimension d, namely

$$d_{G/S} = p^{nd}\mathcal{O}_S,$$

As a corollary, when k is algebraically closed and the fraction field K of R is of characteristic zero, this implies a *formula for the determinant of the Tate module* of G_K,

$$\Lambda^h_{\mathbf{Z}/p^n} G_K \xrightarrow{\sim} (\mathbf{Z}/p^n)(d),$$

where h is the height of G, see ([43, (4.9.2), (4.10)]). These results were used by Raynaud in [65] to effectively bound the modular height in an isogeny class of abelian varieties, an improvement of Faltings' theorem.

(iv) The existence of infinitesimal liftings of BT's and BT$_n$'s was the starting point for the study of their *Dieudonné theory* from a *crystalline* view point. In his letter to Tate [33, (2.6)], Grothendieck makes the following observation. For an abelian scheme $f : A \to S$ over a base S, with dual abelian scheme $A^\vee$, consider the *universal extension* $G(A)$ of A by a vector bundle (a construction due to Serre),

$$0 \to (\mathrm{t}_{A^\vee})^\vee \to G(A) \to A \to 0,$$

where

$$\mathrm{t}_{A^\vee} \xrightarrow{\sim} R^1 f_* \mathcal{O}_A$$

is the Lie algebra of $A^\vee$. The Lie algebra of $G(A)$ is the *dual* $\mathcal{H}^1_{\mathrm{dR}}(A/S)^\vee$ of $\mathcal{H}^1_{\mathrm{dR}}(A/S)$, with its natural filtration:

$$0 \to (\mathrm{t}_{A^\vee})^\vee \to \mathcal{H}^1_{\mathrm{dR}}(A/S)^\vee \to (f_*\Omega^1_{A/S})^\vee \to 0,$$

dual to the Hodge filtration of $\mathcal{H}^1_{\mathrm{dR}}(A/S)$,

$$0 \to f_*\Omega^1_{A/S} \to \mathcal{H}^1_{\mathrm{dR}}(A/S) \to R^1 f_* \mathcal{O}_A \to 0.$$

The crystalline nature of $\mathcal{H}^1_{\mathrm{dR}}$ led Grothendieck to conjecture that the universal extension $G(A)$ itself should be crystalline. Of course, as explained above, the definition of "crystalline" in [33] had to be modified, but with this modified definition, Grothendieck statement was indeed correct. A variant of this extension for BT's (also proposed by Grothendieck in [33, chapter 3]) together with the local liftability statement (4.2.1 (1)) enabled Messing [58] to construct the *Dieudonné crystals* associated to BT's. The theory was developed in several directions afterwards (Mazur-Messing [56], Berthelot-Breen-Messing [7–9]). For a description of the state of the art on this subject in 1998, see de Jong's survey [21]. New breakthroughs were made quite recently by Scholze, using his theory of perfectoid spaces [68], giving in particular a classification of BT's over the ring of integers of an algebraically closed complete extension of $\mathbf{Q}_p$.

Other types of "Dieudonné theories" have been considered. The oldest one is *Cartier's theory* of p-typical curves ([14, 16], see also [53, 75]), which works well for formal groups (even in mixed characteristic). This theory has had a wide range of applications (including K-theory and homotopy theory). For those pertaining to the theory of the *de Rham-Witt complex*, see [44] for a brief survey. More recently, we have the theories of Breuil-Kisin (for finite flat commutative group schemes) ([15,51]), and Zink's theory of *displays* (see Messing's Bourbaki report [59]), which plays an important role in the study of Rapoport-Zink spaces (see (vi) below).

(v) *The mysterious functor, Fontaine's rings and p-adic Hodge theory*. In his talk at the Nice ICM [38] Grothendieck explains that, given a base S where the prime number p is locally nilpotent, and a BT G on S, if $\mathbf{D}(G)$ denotes its Dieudonné crystal (constructed in [58]), the "value" $\mathbf{D}(G)_S$ of $\mathbf{D}(G)$ on S, a locally free $\mathcal{O}_S$-module of rank equal to the height of G, comes equipped with a canonical filtration by a locally direct summand $\mathrm{Fil}(\mathbf{D}(G)_S)$ (namely ω_G), and that if S' is a thickening of S equipped with nilpotent divided powers, then, up to isomorphisms, liftings of G to S' correspond bijectively to liftings of $\mathrm{Fil}(\mathbf{D}(G)_S)$ to a locally direct summand of $\mathbf{D}(G)_{S'}$.

He gives the following corollary. Let R be a complete discrete valuation ring of perfect reside field k of characteristic p and fraction field K of characteristic zero. Let $K_0 := \mathrm{Frac}(W)$, $W = W(k)$. Then the functor associating to a BT G on R *up to isogeny* the pair (M, Fil) consisting of the F-space (see footnote 1) $M = \mathbf{D}(G_k) \otimes_W K_0$ (a K_0-vector space of dimension equal to the height of G, equipped with a σ-linear automorphism F) and the K-*submodule* $\mathrm{Fil} = \mathrm{Fil}\,\mathbf{D}(G_k)_R \otimes_R K \subset M$, is *fully faithful*.

Grothendieck then observes that, on the other hand, in view of Tate's theorem (2.3.1), G is "known" (up to a unique isomorphism) when its Tate module $T_p(G_K)$ is known, therefore raising the question: is there a "more or less algebraic" way of reconstructing (M, Fil) from the datum of the Galois module $T_p(G_K)$? He also proposes to investigate analogues of this question for cohomology in higher degrees, with F-crystals (coming from crystalline cohomology of varieties over k) equipped with longer filtrations (coming from liftings to R). This is the so-called problem of the *mysterious functor*, that he discussed in his talks at the Collège de France (but did not mention explicitly in [38]). As for the original problem (for BT's), in ([31], V 1.4) Fontaine explains how to obtain $T_p(G_K)$ from (M, Fil), but does not define the functor in the other direction. This problem, together with its expected generalizations in higher dimension and the desire to understand its relation with Hodge-Tate's decompositions (2.3.2)), was the starting point of Fontaine's construction of his "Barsotti-Tate rings" (B_{cris}, B_{dR}, B_{st}), and the true beginning of *p-adic Hodge theory*.

(vi) *Rapoport-Zink moduli spaces*. The formal moduli space S in (i) pro-represents the deformation functor of "naked" BT's. In the past fifteen years, variants of this moduli space for BT's endowed with additional structures of isogeny type or complex multiplication type have been constructed by Rapoport-Zink and intensely studied by many other authors. These spaces play the role of local analogues of Shimura varieties arising from moduli of abelian varieties with similar additional structures. They have been used by Harris and Taylor [40] to establish the local Langlands correspondence for GL_n over p-adic fields. A new, simpler proof was recently given by Scholze [66]. See also [30,64,68].

(vii) *Traverso's conjectures.* If G is a BT over an algebraically closed field k of characteristic $p > 0$, there exists a positive integer n such that G is determined up to isomorphism (resp. isogeny) by $G(n)$ (cf. [71, Theorem 3]). The least such n is denoted by n_G (resp. b_G, which is called the *isogeny cutoff* of G). A conjecture of Traverso [72] predicted that the isogeny cutoff of G satisfies the inequality $b_G \leq \lceil dd^\vee/(d+d^\vee)\rceil$ for G of dimension d and codimension $d^\vee$ $(= \dim G^\vee)$, with $dd^\vee > 0$. This conjecture was proved by Nicole and Vasiu [60]. On the other hand, Traverso conjectured that $n_G \leq \min(d, d^\vee)$, but recently, Lau, Nicole and Vasiu [52] disproved this conjecture, giving the correct (sharp) bound $\lfloor 2dd^\vee/(d+d^\vee)\rfloor$. This result makes critical use of 4.2.1. Let me also mention related work of Vasiu [74] and Gabber and Vasiu [32] presenting progress on the search for invariants and classification of truncated BT's.

5. Grothendieck's letter to Barsotti: Newton and Hodge polygons

In 1966-67, during the SGA 6 seminar, Berthelot, Grothendieck and I would often take a walk after lunch in the woods of the I.H.É.S. It is in the course of one of these walks that Grothendieck told us that he had had a look at Manin's paper [55] and thought about his classification theorem (cf. 2.2). What he explained to us that day, he was to write it up years later, in his letter to Barsotti of May 11, 1970 ([39, Appendice]).

Grothendieck observes first that, instead of indexing the simple objects $E_{r,s} = K_\sigma[T]/(T^s - p^r)$ of the category of F-isocrystals on k by pairs of integers (r, s) in lowest terms, it is better to index them by rational numbers, *i.e.* write $E_{r,s} = E_{r/s}$. He calls $\lambda = r/s$ the *slope* of $E_{r,s}$, a terminology which he attributes to Barsotti. In this way, Manin's theorem implies that any F-isocrystal M admits a canonical (finite) decomposition

$$M = \oplus_{\lambda\in\mathbf{Q}} M_\lambda, \qquad (*)$$

where M_λ is *isoclinic* of slope λ, *i.e.* a direct sum of copies of E_λ. This decomposition is compatible with tensor products, and, when k is the algebraic closure of a perfect field k_0, descends to k_0. Ordering the slopes of M in nondecreasing order $\lambda_1 \leq \cdots \leq \lambda_n$ ($n = \mathrm{rk}(M)$), one defines the *Newton polygon* Nwt(M) of M as the graph of the piecewise linear function $0 \mapsto 0, i \in [1, n] \mapsto \lambda_1 + \cdots + \lambda_i$. If m_λ the multiplicity of λ in M (*i.e.* the number of times that λ appears in the preceding sequence), λm_λ is an integer; in particular, the breakpoints lie in $\mathbf{Z}^2$. When k is the algebraic closure of $\mathbf{F}_q, q = p^a$, then (by a result of Manin) this Newton polygon is the Newton polygon of the polynomial $\det(1 - F^a t, M)$. As for the relation with BT's, Grothendieck notes that the (F-iso)crystals corresponding to BT's are those with slopes in the interval [0, 1] (slope 0 (resp. 1) for the ind-étale (resp. multiplicative) ones), and that the decomposition (*) can be refined into a decomposition

$$M = \oplus_{i \in \mathbf{Z}} M_i(-i), \qquad (**)$$

where M_i has slopes in [0, 1) and $(-i)$ is the Tate twist, consisting in replacing F by $p^i F$.

Now, the main points in Grothendieck's letter are:

- a sketch of the proof of a specialization theorem for F-crystals
- a conjecture on the specialization of BT's
- comments on a conjecture of Katz.

I will briefly discuss these points, each of them has had a long posterity.

5.1. The specialization theorem

Roughly speaking, it says that, if M is an absolute F-crystal on a scheme S of characteristic $p > 0$, which one can think of a family of F-crystals M_s parametrized by the points s of S, then the Newton polygon of M_s (*i.e.* of $M_{\bar{s}}$ for a perfect over field $k(\bar{s})$ of $k(s)$) *rises under specialization* of s (and the endpoints don't change). For a more precise statement and a full proof, see ([49, 2.3.1]).

Such F-crystals arise for example from relative crystalline cohomology groups of proper smooth schemes X/S (which, in view of (**), as Grothendieck puts it, produces "a whole avalanche of BT's over k (up to isogeny)"). In this case a variant (and a refinement) of this specialization theorem - which is not a formal consequence of it - was given by Crew [19].

5.2. Specialization of BT's

Grothendieck explains that, in the case of a BT $\mathcal{G}$ over S, of height h and dimension d, with S as in 5.1, the specialization theorem says that, if $G' = \mathcal{G}_s$ is a specialization of $G = \mathcal{G}_t$ ($s \in \overline{\{t\}}$), and (λ_i) (resp. (λ'_i)) $(1 \leq i \leq h)$ is the sequence of slopes of G (resp. G'), then we have

$$\sum \lambda_i = \sum \lambda'_i \tag{1}$$

(both sums being equal to d), and

$$\lambda_1 \leq \lambda'_1,\ \lambda_1 + \lambda_2 \leq \lambda'_1 + \lambda'_2,\ \cdots,\ \sum_{1 \leq i \leq j} \lambda_i \leq \sum_{1 \leq i \leq j} \lambda'_i,\ \cdots \tag{2}$$

((1) expressing that both polygons have the same endpoints $(0, 0), (h, d)$). He conjectures that, conversely, given a BT $G_0 = G'$ over a perfect field k of characteristic $p > 0$, and denoting by S the formal modular variety of G_0 (4.2 (i)), with reduction $S_0 = S \otimes_{W(k)} k$, with universal BT $\mathcal{G}$ over S_0, given any nondecreasing sequence of rational numbers λ_i $(1 \leq i \leq h)$ between 0 and 1, the conditions (1) and (2) are sufficient for the existence of a fiber of $\mathcal{G}$ at some point of S_0 having this sequence as sequence of slopes. This conjecture was eventually proven by Oort in 2000 [63].

5.3. Katz's conjecture

At the end of his letter, Grothendieck says that his specialization theorem was suggested to him by "a beautiful conjecture of Katz", which he recalls and formulates in a greater generality. This is the following statement:

Conjecture 5.3.1. Let k be a perfect field of characteristic $p > 0$, $W = W(k)$, $K_0 = \mathrm{Frac}(W)$, X/k a proper and smooth scheme, $i \in \mathbf{Z}$, $H^i(X/W)$ the i-th crystalline cohomology group of X/k, with its σ-linear endomorphism F. Let $\mathrm{Nwt}_i(X)$ be the Newton polygon of the F-isocrystal $(H^i(X/W) \otimes K_0, F)$. Let $\mathrm{Hdg}_i(X)$ be the *Hodge polygon* of $H^i_{\mathrm{Hdg}}(X/k) = \oplus H^{i-j}(X, \Omega^j_{X/k})$, starting at 0 and having slope j with multiplicity $h^{j,i-j} = \dim_k H^{i-j}(X, \Omega^j_{X/k})$. Then $\mathrm{Nwt}_i(X)$ lies on or above $\mathrm{Hdg}_i(X)$.

As recalled in ([50, page 343]), such an inequality was proved for the first time by Dwork, for the middle dimensional primitive cohomology of a projective smooth hypersurface of degree prime to p ([25, Section 6]). Conjecture 5.3.1 was established first by Mazur for X liftable to W, and then by Ogus in general ([10, Section 8]), with a refinement when X/k has nice cohomological properties, namely that $H^*(X/W)$ is torsion-free, and the Hodge to de Rham spectral sequence of X/k degenerates at E_1. See [44] for a survey.

Grothendieck adds that in the case where X/k "lifts to characteristic zero", one should have a stronger inequality, involving the Hodge numbers of the lifted variety. Namely, if X'/R is a proper and smooth scheme over R, R a complete discrete valuation ring with residue field k and fraction field K of characteristic 0, such that $X' \otimes k = X$, then one can consider the Hodge numbers $h'^{j,i-j} = \dim_k H^{i-j}(X'_K, \Omega^j_{X'_K/K})$, which satisfy

$$h'^{0,i} \leq h^{0,i}, \cdots, h'^{j,i-j} \leq h^{j,i-j}, \cdots,$$

so that the Hodge polygon $\mathrm{Hdg}_i(X'_K)$, constructed similarly to $\mathrm{Hdg}_i(X)$ but with the numbers $h'^{j,i-j}$, lies on or above $\mathrm{Hdg}_i(X)$. Then he proposes:

Conjecture 5.3.2. With the above notations, $\mathrm{Nwt}_i(X/k)$ lies on or above $\mathrm{Hdg}_i(X'_K)$.

Grothendieck says that he has some idea on how to attack 5.3.1, but not 5.3.2 "for the time being". Actually, 5.3.2 was to follow from the proof of Fontaine's conjecture C_{cris}, which implies that the filtered φ-module

$$((H^i(X_k/W), F), \mathrm{Fil}^j H^i_{\mathrm{dR}}(X'_K/K)),$$

where Fil^j denotes the Hodge filtration, is *weakly admissible*.

Inequalities 5.3.1 and 5.3.2 have applications to Chevalley-Warning type congruences on numbers of rational points of varieties over finite fields (or over discrete valuation rings R as above with k finite). See [44] and [45] for 5.3.1. As an example of application of 5.3.2, quite recently Berthelot, Esnault and Rülling used a variant of 5.3.2 (for proper flat schemes having *semistable reduction* over k), following from the proof of Fontaine-Jannsen's C_{st}-conjecture, together with several other cohomological techniques (Berthelot's *rigid cohomology*, Witt vectors cohomology) to prove the following theorem:

Theorem 5.3.3. (**[12]**) *Let X/R with R as above and $k = \mathbf{F}_q$. Assume:*

(i) *X regular, and proper and flat over R;*
(ii) *X_K geometrically connected;*
(iii) *$H^i(X_K, \mathcal{O}_{X_K}) = 0$ for all $i > 0$.*

Then $|X_k(\mathbf{F}_{q^n})| \equiv 1 \bmod q^n$ for all $n \geq 1$.

See the introduction of [12] for a discussion of the analogy of this result with that of Esnault [26] based on ℓ-adic techniques.

5.4. New viewpoints on slopes

The analogy between the notions of slopes and Newton polygons for F-crystals and those of slopes and Harder-Narasimhan filtrations for vector bundles on curves is not an accident. There is a common framework for the two notions, which was recently discovered by André [1]. Fargues [28] exploited this to construct a Harder-Narasimhan filtration on finite flat commutative group schemes over valuation rings of mixed characteristics, and similar filtrations play an important role in Fargues-Fontaine's work [29] on p-adic Galois representations.

ACKNOWLEDGEMENTS. I wish to thank Jean-Marc Fontaine, Nick Katz, Bill Messing, Marc-Hubert Nicole, Peter Scholze, Jean-Pierre Serre, and Weizhe Zheng for helpful comments on preliminary versions of this text.

References

[1] Y. ANDRÉ, *Slope filtrations*, Confluentes Mathematici **1** (2009), 1–85.

[2] I. BARSOTTI, *Analytical methods for abelian varieties in positive characteristic*, Colloq. Théorie des Groupes Algébriques (Bruxelles, 1962), Librairie Universitaire, Louvain, 77–85.

[3] A. BEILINSON, *p-adic periods and de Rham cohomology*, J. of the AMS **25** (2012), 715–738.

[4] A. BEILINSON, *On the crystalline period map*, preprint (2013), http://arxiv.org/abs/1111.3316v4.

[5] P. BERTHELOT, "Cohomologie Cristalline des Schémas de Caractéristique $p > 0$", Lecture Notes in Math., Vol. 407, Springer-Verlag, 1974.

[6] P. BERTHELOT, *Altérations de variétés algébriques*, Séminaire N. Bourbaki, 1995-1996, n. 815, 273-311, Astérisque 241, 1997.

[7] P. BERTHELOT and W. MESSING, *Théorie de Dieudonné cristalline I*, in Journées de Géométrie Algébrique de Rennes I, Astérisque 63, 1979, 17–37.

[8] P. BERTHELOT, L. BREEN and W. MESSING, "Théorie de Dieudonné Cristalline II", Lecture Notes in Math., Vol. 930, Springer-Verlag, 1982.

[9] P. BERTHELOT and W. MESSING, *Théorie de Dieudonné Cristallin III: Théorèmes d'Équivalence et de Pleine Fidélité*, in The Grothendieck Festschrift Vol. I, P. Cartier, L. Illusie, N.M. Katz, G. Laumon, Y. Manin and K.A. Ribet (eds.), Progress in Math., Vol. 86, Birkhäuser, 1990, 173–247.

[10] P. BERTHELOT and A. OGUS, "Notes on Crystalline Cohomology", Math. Notes, Vol. 21, Princeton Univ. Press, Princeton, NJ, 1978.

[11] P. BERTHELOT and A. OGUS, *F-Isocrystals and De Rham Cohomology. I*, Invent. Math. **72** (1983), 159–199.

[12] P. BERTHELOT, H. ESNAULT and K. RÜLLING, *Rational points over finite fields for regular models of algebraic varieties of Hodge type* ≥ 1, Ann. of Math. **176**, 1 (2012), 413–508, arXiv 1009.0178.

[13] L. BREEN, *Extensions du groupe additif*, Pub. Math. I.H.É.S. **48** (1978), 39–125.

[14] P. CARTIER, *Groupes formels associés aux anneaux de Witt généralisés*, C. R. Acad. Sci. Paris Sér. A-B **265** (1967), A49–A52.

[15] C. BREUIL, *Groupes p-divisibles, groupes finis et modules filtrés*, Ann. of Math. (2) **152** (2000), no. 2, 489–549.

[16] P. CARTIER, *Modules associés à un groupe formel commutatif. Courbes typiques*, C. R. Acad. Sci. Paris Sér. A-B **265** (1967), A129–A132.

[17] P. CARTIER, "Relèvements des groupes formels commutatifs", Séminaire Bourbaki, 1968-69, 359, Lecture Notes in Math., Vol. 179, Springer-Verlag, 1972, 217–230.

[18] P. COLMEZ et J.-P. SERRE (eds.) "Correspondance Grothendieck-Serre", Documents Mathématiques 2, SMF, 2001.

[19] R. CREW, *Specialization of crystalline cohomology*, Duke Math. J. **53** (1986), 749–757.

[20] A. J. DE JONG, *Smoothness, semi-stability and alterations*, Pub. math. I.H.É.S. **83** (1996), 51–93.

[21] J. DE JONG, *Barsotti-Tate groups and crystals*, Proceedings of the International Congress of Mathematicians, Vol. II (Berlin, 1998), Documenta Mathematica II: 259–265.

[22] P. DELIGNE, Letter to L. Breen, Aug. 4, 1970 (unpublished).

[23] M. DEMAZURE, "Lectures on p-divisible Groups", Lecture Notes in Mathematics, Vol. 302, Springer-Verlag, 1972.

[24] B. DWORK, *On the rationality of the zeta function of an algebraic variety*, Amer. J. Math. **82** (1960), 631–648.

[25] B. DWORK, *A deformation theory for the zeta function of a hypersurface*, Proc. ICM Stockholm (1962), 247–259.

[26] H. ESNAULT, *Deligne's integrality theorem in unequal characteristic and rational points over finite fields*, Annals of Math. **164** (2006), 715–730.

[27] G. FALTINGS, *Almost étale extensions*, Cohomologies p-adiques et applications arithmétiques (II), Astérisque 279, SMF, 2002, 185–270.

[28] L. FARGUES, *La filtration de Harder-Narasimhan des schémas en groupes finis et plats*, J. für die reine und angewandte Mathematik (Crelle J.) **645** (2010), 1–39.

[29] L. FARGUES and J.-M. FONTAINE, *Vector bundles and p-adic Galois representations*, AMS/IP Studies in Advanced Math. **51**, 5th International Congress of Chinese Mathematicians, AMS 2012.

[30] L. FARGUES and E. MANTOVAN, *Variétés de Shimura, espaces de Rapoport-Zink et correspondances de Langlands locales* Astérisque 291, SMF, 2004.

[31] J.-M. FONTAINE, *Groupes p-divisibles sur les corps locaux*, Astérisque 47–48, SMF, 1977.

[32] O. GABBER and A. VASIU, *Dimensions of group schemes of automorphisms of truncated Barsotti-Tate groups*, IMRN (2013) 2013 (18): 4285–4333. DOI: 10.1093/imrn/rns165.

[33] A. GROTHENDIECK, Letter to J. Tate, May 1966.

[34] A. GROTHENDIECK, *On the de Rham cohomology of algebraic varieties*, Pub. math. I.H.É.S. **29** (1966), 95–103.

[35] A. GROTHENDIECK, *Crystals and the de Rham cohomology of schemes* (I.H.É.S., Dec. 1966), notes by J. Coates and O. Jussila, in *Dix exposés sur la cohomologie des schémas*, Advanced Studies in Pure Math. 3, North-Holland, Masson, 1968.

[36] A. GROTHENDIECK, Letter to L. Illusie, 2-4 Dec. 1969.

[37] A. GROTHENDIECK, "Groupes de Monodromie en Géométrie Algébrique", Séminaire de Géométrie Algébrique du Bois-Marie 1967-69 (SGA 7 I), Lecture Notes in Math., Vol. 288, Springer-Verlag, 1972.

[38] A. GROTHENDIECK, *Groupes de Barsotti-Tate et cristaux*, Actes du Congrès International des Mathématiciens, Nice (1970), Gauthier-Villars, 1971, 431–436.

[39] A. GROTHENDIECK, *Groupes de Barsotti-Tate et cristaux de Dieudonné*, Sém. Math. Sup. 45, Presses de l'Université de Montréal, 1974.

[40] M. HARRIS and R. TAYLOR, *The geometry and cohomology of some simple Shimura varieties*, Annals of Mathematics Studies, Princeton University Press, 2001.

[41] L. ILLUSIE, "Complexe Cotangent et Déformations II", Lecture Notes in Math., Vol. 283, Springer-Verlag, 1972.

[42] L. ILLUSIE, *Complexe de de Rham-Witt et cohomologie cristalline*, Ann. Sci. ÉNS, 4e série, t. 12 (1979), 501–661.

[43] L. ILLUSIE, *Déformations de groupes de Barsotti-Tate, d'après A. Grothendieck*, In: "Séminaire sur les pinceaux arithmétiques :

la conjecture de Mordell, Lucien Szpiro", Astérisque 127, SMF, 1985, 151–198.

[44] L. ILLUSIE, *Crystalline cohomology*, In: "Motives", U. Jannsen, S. Kleiman and J.-P. Serre (eds.), Proceedings of Symposia in Pure Math. 55, 1, 43–70.

[45] L. ILLUSIE, *Miscellany on traces in ℓ-adic cohomology: a survey*, Japanese Journal of Math. **1-1** (2006), 107–136.

[46] N. KATZ and T. ODA, *On the differentiation of De Rham cohomology classes with respect to parameters*, J. Math. Kyoto Univ. Volume 8, Number 2 (1968), 199–213.

[47] N. KATZ, *Nilpotent connections and the monodromy theorem: applications of a result of Turrittin*, Pub. Math. I.H.É.S. **39** (1970), 175–232.

[48] N. KATZ, *Serre-Tate local moduli*, Surfaces Algébriques, Séminaire de Géométrie Algébrique d'Orsay 1976-78, Lecture Notes in Math., Vol. 868, Springer-Verlag, 1981, 138-202.

[49] N. KATZ, *Slope filtration of F-crystals*, in Journées de Géométrie Algébrique de Rennes I, Astérisque 63, SMF, 1979, 113–163.

[50] N. KATZ and J. TATE, *Bernard Dwork (1923-1998)*, Notices of the AMS **46**, n.3 (1999), 338–343.

[51] M. KISIN, *Crystalline representations and F-crystals*, Algebraic geometry and number theory, 459–496, Progr. Math., 253, Birkhäuser Boston, Boston, MA, 2006.

[52] E. LAU, M.-H. NICOLE and A. VASIU, *Stratification of Newton polygon strata and Traverso's conjectures for p-divisible groups*, Ann. of Math. **178**, 3 (2013), 789–834.

[53] M. LAZARD, "Commutative Formal Groups", Lecture Notes in Math., Vol. 443, Springer-Verlag, 1975.

[54] J. LUBIN, J-P. SERRE and J. TATE, *Elliptic curves and formal groups*, Woods Hole Summer Institute, mimeographed notes, 1964.

[55] Y. MANIN, *The theory of commutative formal groups over fields of finite characteristic*, English trans., Russian Math. Surv. **18**, 1963.

[56] B. MAZUR and W. MESSING, "Universal Extensions and One Dimensional Crystalline Cohomology", Lecture Notes in Math., Vol. 370, Springer-Verlag, 1974.

[57] B. MAZUR and L. ROBERTS, *Local Euler characteristics*, Invent. Math. **2** (1970), 201–234.

[58] W. MESSING, "The Crystals Associated to Barsotti-Tate Groups: with Applications to Abelian Schemes", Lecture Notes in Math., Vol. 264, Springer-Verlag, 1972.

[59] W. MESSING, *Travaux de Zink*, Séminaire Bourbaki, 2005-2006, n. 964, Astérisque 311, SMF, 2007, 341–364.

[60] M.-H. NICOLE and A. VASIU, *Traverso's Isogeny Conjecture for p-Divisible Groups*, Rend. Sem. Univ. Padova, **118** (2007), 73–83.

[61] W. NIZIOL, *Semistable conjecture via K-theory*, Duke Math. J. **141** (2008), 151–178.

[62] T. ODA, *The First de Rham Cohomology Group and Dieudonné Modules*, Ann. Scient. ÉNS, 3e série, t. 2 (1969), 63–135.

[63] F. OORT, *Newton polygons and formal groups: conjectures by Manin and Grothendieck*, Ann. of Math. **152** (2000), 183–206.

[64] M. RAPOPORT and T. ZINK, "Period Spaces for p-divisible Groups", Ann. Math. Stud., Vol. 141, Princeton University Press, Princeton 1996.

[65] M. RAYNAUD, *Hauteurs et isogénies*, In: "Séminaire sur les pinceaux arithmétiques : la conjecture de Mordell", Lucien Szpiro, Astérisque 127, SMF, 1985, 199–234.

[66] P. SCHOLZE, *The Local Langlands Correspondence for* GL_n *over p-adic fields*, Invent. Math. 192, 3 (2013), 663–715.

[67] P. SCHOLZE, *p-adic Hodge theory for rigid-analytic varieties*, arXiv:1205.3463v2 [math. AG] (2012).

[68] P. SCHOLZE and J. WEINSTEIN, *Moduli of p-divisible groups*, arXiv:1211.6357v2 [math. NT] (2013).

[69] J-P. SERRE, *Groupes p-divisibles (d'après J. Tate)*, Séminaire Bourbaki, 1966/67, n. 318, in Réédition du Séminaire Bourbaki 1948-1968, volume 10, Années 1966-67, 1967-68, SMF.

[70] J. TATE, *p-Divisible Groups*, Proc. of a Conference on Local Fields, Driebergen 1966 (Ed. T. A. Springer), 158–183, Springer-Verlag, 1987.

[71] C. TRAVERSO, *Sulla classificazione dei gruppi analitici di caratteristica positiva*, Ann. Scuola Norm. Sup. Pisa **23**, 3 (1969), 481–507.

[72] C. TRAVERSO, *Specialization of Barsotti-Tate groups*, Symposia Mathematica, Vol. XXIV (Sympos., INDAM, Rome, 1979), 1–21, Academic Press, 1981.

[73] T. TSUJI, *p-adic étale cohomology and crystalline cohomology in the semi-stable reduction case*, Invent. Math. **137** (1999), 233–411.

[74] A. VASIU, *Reconstructing p-divisible groups from their truncations of small level*, Comm. Math. Helv. **85** (2010), 165–202.

[75] T. ZINK, *Cartiertheorie kommutativer former Gruppen*, Teubner, 1984.

[76] [SGA 4] *Théorie des topos et cohomologie étale des schémas*, Séminaire de géométrie algébrique du Bois-Marie 1963-64, dirigé par M. Artin, A. Grothendieck, J.-L. Verdier, Lecture Notes in Math., Vols. 269, 270, 305, Springer-Verlag, 1972, 1973.

[77] [SGA 5] *Cohomologie ℓ-adique et fonctions L*, Séminaire de Géométrie Algébrique du Bois-Mari.e 1965/66, dirigé par A. Grothendieck, Lecture Notes in Math., Vol. 589, Springer-Verlag, 1977.

[78] [SGA 6] *Théorie des Intersections et Théorème de Riemann-Roch*, Séminaire de géométrie algébrique du Bois-Marie 1966-67, dirigé par P. Berthelot, A. Grothendieck, L. Illusie, Lecture Notes in Math., Vol. 225, Springer-Verlag, 1971.

[79] [SGA7] *Groupes de monodromie en géométrie algébrique*, Séminaire de Géométrie Algébrique du Bois-Marie 1967-1969, I dirigé par A. Grothendieck, II par P. Deligne et N. Katz, Lecture Notes in Math., Vols. 288, 340, Springer-Verlag 1972, 1973.

Riemann's hypothesis

Brian Conrey

1. Gauss

There are 4 prime numbers less than 10; there are 25 primes less than 100; there are 168 primes less than 1000, and 1229 primes less than 10000. At what rate do the primes thin out? Today we use the notation $\pi(x)$ to denote the number of primes less than or equal to x; so $\pi(1000) = 168$.

Carl Friedrich Gauss in an 1849 letter to his former student Encke provided us with the answer to this question. Gauss described his work from 58 years earlier (when he was 15 or 16) where he came to the conclusion that the likelihood of a number n being prime, without knowing anything about it except its size, is

$$\frac{1}{\log n}.$$

Since $\log 10 = 2.303\ldots$ the means that about 1 in 16 seven digit numbers are prime and the 100 digit primes are spaced apart by about 230. Gauss came to his conclusion empirically: he kept statistics on how many primes there are in each sequence of 100 numbers all the way up to 3 million or so! He claimed that he could count the primes in a chiliad (a block of 1000) in 15 minutes! Thus we expect that

$$\pi(N) \approx \frac{1}{\log 2} + \frac{1}{\log 3} + \frac{1}{\log 4} + \cdots + \frac{1}{\log N}.$$

This is within a constant of

$$\mathrm{li}(N) = \int_0^N \frac{du}{\log u}$$

so Gauss' conjecture may be expressed as

$$\pi(N) = \mathrm{li}(N) + E(N)$$

where $E(N)$ is an error term. What is stunning is how small $E(N)$ is! Here is some data from Gauss' letter:

N	$\pi(N)$	$\mathrm{li}(N)$	$E(N)$
500000	41556	41606.4	+50.4
1000000	78501	78627.5	+126.5
1500000	114112	114263.1	+151.1
2000000	148883	149054.8	+171.8
2500000	183016	183245.0	+229.0
3000000	216745	216970.6	+225.6

Note that there is an asymptotic expansion

$$\mathrm{li}(N) = \frac{N}{\log N} + \frac{N}{\log^2 N} + \dots.$$

2. Riemann

In 1859 G. B. F. Riemann proposed a pathway to understand on a large scale the distribution of the prime numbers. He studied

$$\zeta(s) = 1 + \frac{1}{2^s} + \frac{1}{3^s} + \dots$$

as a function of a complex variable $s = \sigma + it$ and proved that $\zeta(s)$ has a meromorphic continuation to the whole s – plane with its only singularity a simple pole at $s = 1$. Moreover, he proved a functional equation that relates $\zeta(s)$ to $\zeta(1-s)$ in a simple way and reveals that $\zeta(-2) = \zeta(-4) = \dots = 0$. His famous hypothesis is that all the other zeros have real part equal to 1/2. He even computed the first few of these non-real zeros:

$$1/2 + i14.13\dots, /2 + i21.02\dots, 1/2 + i25.01\dots, \dots$$

A good way to be convinced that these are indeed zeros is to use the easily proven formula

$$(1 - 2^{1-s})\zeta(s) = 1 - \frac{1}{2^s} + \frac{1}{3^s} - \frac{1}{4^s} \pm \dots.$$

The alternating series on the right converges for $\Re s > 0$ and so

$$s = 1/2 + i14.134725141734693790457251983562 4\dots$$

can be substituted into a truncation of this series (using your favorite computer algebra system) to see that it is very close to 0. (See www.lmfdb.org

to find a list of high precision zeros of $\zeta(s)$ as well as a wealth of information about $\zeta(s)$ and similar functions called L-functions.)

Euler had earlier proved that $\zeta(s)$ can be expressed as an infinite product over primes:

$$\zeta(s) = \left(1+\frac{1}{2^s}+\frac{1}{4^s}+\frac{1}{8^s}+\dots\right)\left(1+\frac{1}{3^s}+\frac{1}{9^s}+\frac{1}{27^s}+\dots\right)\left(1+\frac{1}{5^s}+\dots\right)$$
$$= \prod_p (1-p^{-s})^{-1}.$$

Euler's formula is essentially an analytic encoding of the fundamental theorem of arithmetic that each integer can be expressed uniquely as a product of primes. Euler's formula provided a clue for Riemann to use complex analysis to investigate the primes. We can almost immediately see a consequence of Riemann's Hypothesis about the zeros of $\zeta(s)$ if we invert Euler's formula:

$$\frac{1}{\zeta(s)} = \prod_p (1-p^{-s}) = 1 - 2^{-s} - 3^{-s} - 5^{-s} + 6^{-s} + \cdots = \sum_{n=1}^{\infty} \frac{\mu(n)}{n^s}$$

where μ is known as the Möbius mu-function. A simple way to explain the value of $\mu(n)$ is that it is 0 if n is divisible by the square of any prime, while if n is squarefree then it is $+1$ if n has an even number of prime divisors and -1 if n has an odd number of prime divisors. Not surprisingly the series

$$\sum_{n=1}^{\infty} \frac{\mu(n)}{n^s}$$

converges if the real part of s is bigger than the real part of any zero ρ of $\zeta(s)$. This actually translates into an equivalent formulation of the Riemann Hypothesis (RH):

$$\text{RH is true if and only if for any } \epsilon > 0, \sum_{n \le x} \mu(n) \le C(\epsilon) x^{1/2+\epsilon}.$$

Thus, we expect that the integers with an even number of prime factors are equally numerous as the integers with an odd number of prime factors and the difference between the two counts is related to the size of the rightmost zero of $\zeta(s)$.

Riemann had an ambitious plan to find an *exact* formula for the number of primes up to x and he did so:

$$\pi(x) = \sum_{n=1}^{\infty} \frac{\mu(n)}{n} f(x^{1/n})$$

where

$$f(x) = \mathrm{li}(x) - \sum_{\rho} \mathrm{li}(x^{\rho}) - \ln 2 + \int_x^{\infty} \frac{dt}{t(t^2-1)\log t}.$$

Here the $\rho = \beta + i\gamma$ are the zeros of $\zeta(s)$. The upshot of all of this is that the error term $E(x)$ from Gauss' conjecture is no more than $x^{\beta_0} \log x$ where β_0 is the supremum of the real parts of the zeros $\rho = \beta + i\gamma$ of $\zeta(s)$. So, if Riemann's hypothesis is true then Gauss' formula is correct with a square-root sized error term! Whereas even one zero not on the 1/2–line will lead to a an error term larger by a power of x. Note that by Riemann's functional equation the zeros are symmetric with respect to the 1/2-line. And $\zeta(s)$ is real for real s so the zeros are symmetric with respect to the real axis. Thus, if $\rho = \beta + i\gamma$ is a zero then so are $\beta - i\gamma$ and if $\beta \neq 1/2$ then $1 - \beta \pm i\gamma$ are zeros, too.

There are no zeros with real part greater than 1; we know this because of Euler's product formula. But if a zero had a real part equal to 1 then the error term in Gauss formula would be as large as the main term! So, it was a huge advance in 1896 when Hadamard and de la Vallée Poussin independently proved that $\zeta(1 + it) \neq 0$ and concluded that

$$\pi(N) \sim \mathrm{li}(N)$$

a theorem which is known as the prime number theorem.

3. How many zeros are there

The non-trivial zeros $\rho = \beta + i\gamma$ all satisfy $0 < \beta < 1$ and Riemann's Hypothesis is that $\beta = 1/2$ always. It has been proven that each of the first 10 trillion zeros have real part equal to 1/2! This is very compelling evidence. Riemann gave us an accurate count of the zeros:

$$N(T) := \#\{\rho = \beta + i\gamma | 0 < \gamma \leq T\} = \frac{T}{2\pi} \log \frac{T}{2\pi e} + O(\log T).$$

Thus, on average there are $\frac{\log T}{2\pi}$ zeros with γ between T and $T + 1$. So the zeros are becoming denser as we go up the critical strip $0 < \sigma < 1$. Hardy was the first person to prove that infinitely many of the zeros are on the 1/2-line! If we let

$$N_0(T) := \#\{\rho = \beta + i\gamma | \beta = 1/2 \text{ and } 0 < \gamma \leq T\}$$

then Hardy and Littlewood proved that

$$N_0(T) \geq CT$$

for some constant $C > 0$. In 1946 Selberg showed that

$$N_0(T) \geq CN(T)$$

for a $C > 0$, *i.e.* that a positive proportion of zeros are on the critical line $\sigma = 1/2$. Levinson [5], in 1974, by a method different than Selberg's showed that $C = 1/3$ is admissible. Conrey [2] in 1987 proved that $C = 0.4088$ is allowable; the current record is due to Feng: $C = 0.412$.

There are also density results. Roughly speaking these say that almost all of the zeros of $\zeta(s)$ are very near to the critical line $\sigma = 1/2$. A classical theorem (see [8, Section 9.19]) is that

$$N(\sigma, T) := \#\{\rho = \beta + i\gamma : \beta \geq \sigma \text{ and } \gamma \leq T\} < CT^{(3/2-\sigma)} \log^5 T.$$

In particular, for any $\sigma > 1/2$ the number of zeros to the right of the σ line with imaginary parts at most T is bounded by T to a power smaller than 1 and so is $o(T)$.

4. Approaches to RH

The Riemann zeta-function $\zeta(s)$ is the last elementary function that we do not understand. There are numerous approaches to RH and hundreds of equivalent formulations in nearly every field of mathematics. But, at the moment there is no clear trail; RH stands alone as a singular monument. It is generally regarded as the most important unsolved problem in all of mathematics. It is pretty universally believed to be true. There are near counter examples and many many wrong proofs. These usually involve Dirichlet series that satisfy functional equations similar to that of $\zeta(s)$ but do not have an Euler product. The belief is that one must use these two ingredients appropriately to make any progress. In this sense, one might say that RH is a fundamental statement about a relationship between addition and multiplication that we still do not understand.

4.1. Almost periodicity

A tempting strategy is to try to prove that if $\zeta(s)$ has one zero off the line then it has infinitely many off the line. Bombieri has come closest to achieving this. One reason this is tempting is the analogy with almost periodic functions in the sense of Bohr. Dirichlet polynomials $\sum_{n=1}^{N} a_n n^{-s}$ are almost periodic. They have the property that a zero 'repeats itself' approximately. If you find one zero then moving vertically in the s-plane one finds $> CT$ zeros up to a height T. If the zeta-function were almost periodic in the half plane $\sigma > 1/2$ then one would expect that a zero $\beta + i\gamma$ with $\beta > 1/2$ off the line would lead to CT zeros off the line in some half plane $\sigma \geq \sigma_0 > 1/2$. But this would contradict the density theorem mentioned above. Here is a conjecture that might encapsulate this idea:

Conjecture 4.1. Suppose that the Dirichlet series

$$F(s) = \sum_{n=1}^{\infty} a_n n^{-s}$$

converges for $\sigma > 0$ and has a zero in the half-plane $\sigma > 1/2$. Then there is a number $C_F > 0$ such that $F(s)$ has $> C_F T$ zeros in $\sigma > 1/2$, $|t| \leq T$.

This seemingly innocent conjecture implies the Riemann Hypothesis for virtually any primitive L-function (except curiously possibly the Riemann zeta-function itself!). And it seems that the Euler product condition has already been used (in the density result above); *i.e.* the hard part is already done. Note that the "1/2" in the conjecture needs to be there as the example

$$\sum_{n=1}^{\infty} \frac{\mu(n)/n^{1/2}}{n^s}$$

demonstrates. Assuming RH this series converges for $\sigma > 0$ and its lone zero is at $s = 1/2$. This example is possibly at the boundary of what is possible.

5. A spectral interpretation

Hilbert and Pólya are reputed to have suggested that the zeros of $\zeta(s)$ should be interpreted as eigenvalues of an appropriate operator.

In the 1950s physicists predicted that excited nuclear particles emit energy at levels which are distributed like the eigenvalues of random matrices. This was verified experimentally in the 1970s and 1980s; Oriol Bohigas was the first to put this data together in a way that demonstrated this law.

In 1972 Hugh Montgomery, then a graduate student at Cambridge, delivered a lecture at a symposium on analytic number theory in St. Louis, outlining his work on the spacings between zeros of the Riemann zeta-function. This was the first time anyone had considered such a question. On his flight back to Cambridge he stopped over in Princeton to show his work to Selberg. At afternoon tea at the Institute for Advanced Study, Chowla insisted that Montgomery meet the famous physicist – and former number theorist – Freeman Dyson. When Montgomery explained to Dyson the kernel he had found that seemed to govern the spacings of pairs of zeros, Dyson immediately responded that it was the same kernel that governs pairs of eigenvalues of random matrices.

In 1980, Andrew Odlyzko and Schonhage invented an algorithm which allowed for the very speedy calculation of many values of $\zeta(s)$ at once. This led Odlyzko to do compile extensive statistics about the zeros at enormous heights - up to 10^{20} and higher. His famous graphs showed an incredible match between data for zeros of $\zeta(s)$ and for the proven statistical distributions for random matrices.

These amazing graphs reminded people of the Pólya and Hilbert philosophy and prompted Odlyzko to write to Pólya. Here is the text of Odlyzko's letter, dated Dec. 8, 1981.

> Dear Professor Pólya:
> I have heard on several occasions that you and Hilbert had independently conjectured that the zeros of the Riemann zeta function correspond to the eigenvalues of a self-adjoint hermitian operator. Could you provide me with any references? Could you also tell me when this conjecture was made, and what was your reasoning behind this conjecture at that time?
> The reason for my questions is that I am planning to write a survey paper on the distribution of zeros of the zeta function. In addition to some theoretical results, I have performed extensive computations of the zeros of the zeta function, comparing their distribution to that of random hermitian matrices, which have been studied very seriously by physicists. If a hermitian operator associated to the zeta function exists, then in some respects we might expect it to behave like a random hermitian operator, which in turn ught to resemble a random hermitian matrix. I have discovered that the distribution of zeros of the zeta function does indeed resemble the distribution of eigenvalues of random hermitian matrices of unitary type.
> Any information or comments you might care to provide would be greatly appreciated.
>
> Sincerely yours,
> Andrew Odlyzko

and Pólya's response, dated January 3, 1982.

> Dear Mr. Odlyzko,
> Many thanks for your letter of Dec. 8. I can only tell you what happened to me.
> I spent two years in Göttingen ending around the beginning of 1914. I tried to learn analytic number theory from Landau. He asked me one day: "You know some physics. Do you know a physical reason that the Riemann Hypothesis should be true?"

This would be the case, I answered, if the non-trivial zeros of the ζ function were so connected with the physical problem that the Riemann Hypothesis would be equivalent to the fact that all the eigenvalues of the physical problem are real.
I never published this remark, but somehow it became known and it is still remembered.
With best regards.

Yours sincerely,
George, Pólya

Now we have the challenge of not only explaining why all of the zeros are on a straight line, but also why they are distributed on this line the way they are! The connections with Random Matrix theory first discovered by Montgomery and Dyson have received a great deal of support from seminal papers of Katz and Sarnak and Keating and Snaith. The last 15 years have seen an explosion of work around these ideas. In particular, it definitely seems like there should be a spectral interpretation of the zeros à la Hilbert and Pólya.

References

[1] E. BOMBIERI, *Remarks on Weil's quadratic functional in the theory of prime numbers. I*, Atti Accad. Naz. Lincei Cl. Sci. Fis. Mat. Natur. Rend. Lincei (9) Mat. Appl. **11** (2000), no. 3, 183–233 (2001).

[2] J. B. CONREY, *More than two fifths of the zeros of the Riemann zeta function are on the critical line*, J. Reine Angew. Math. **399** (1989), 1–26.

[3] N. M. KATZ and P. SARNAK, "Random Matrices", Frobenius eigenvalues, and monodromy, American Mathematical Society Colloquium Publications, 45, American Mathematical Society, Providence, RI, 1999.

[4] J. P. KEATING and N. C. SNAITH, *Random matrix theory and* $\zeta(1/2+it)$, Comm. Math. Phys. **214** (2000), no. 1, 57–89.

[5] N. LEVINSON, *More than one third of zeros of Riemann's zeta-function are on* $\sigma = 1/2$, Advances in Math. **13** (1974), 383–436.

[6] M. L. MEHTA, "Random Matrices", Second edition, Academic Press, Inc., Boston, MA, 1991.

[7] H. L. MONTGOMERY, *The pair correlation of zeros of the zeta function*, In: "Analytic Number Theory", Proc. Sympos. Pure Math., Vol. XXIV, St. Louis Univ., St. Louis, Mo., 1972, pp. 181–193. Amer. Math. Soc., Providence, R.I., 1973.

[8] E. C. TITCHMARSH, "The Theory of the Riemann Zeta-Function", Second edition. Edited and with a preface by D. R. Heath-Brown. The Clarendon Press, Oxford University Press, New York, 1986.

[9] A. M. ODLYZKO, *On the distribution of spacings between zeros of the zeta function*, Math. Comp. **48** (1987), no. 177, 273–308.

Stability results for the Brunn-Minkowski inequality

Alessio Figalli

1. Introduction

The Brunn-Miknowski inequality gives a lower bound on the Lebesgue measure of a sumset in terms of the measures of the individual sets. This classical inequality in convex geometry was inspired by issues around the isoperimetric problem and was considered for a long time to belong to geometry, where its significance is widely recognized. However, it is by now clear that the Brunn-Miknowski inequality has also applications in analysis, statistics, informations theory, etc. (we refer the reader to [29] for an extended exposition on the Brunn-Minkowski inequality and its relation to several other famous inequalities).

To focus more on the analytic side, we recall that Brunn-Minkowski **(BM)** is intimately connected to several other famous inequalities such as the isoperimetric **(Isop)** inequality, Sobolev **(Sob)** inequalities, and Gagliardo-Nirenberg **(GN)** inequalities. In particular, it is well-known that the following chain of implications holds, although in general one cannot obtain one inequality from the other with sharp constants (see for instance [20] for a more detailed discussion):

$$\textbf{(BM)} \quad \Rightarrow \quad \textbf{(Isop)} \quad \Rightarrow \quad \textbf{(Sob)} \quad \Rightarrow \quad \textbf{(GN)}.$$

The issue of the sharpness of a constant, as well as the characterization of minimizers, is a classical and important question which is by now well understood (at least for the class of inequalities we are considering). More recently, a lot of attention has been given to the stability issue:

Suppose that a function almost attains the equality in one of the previous inequalities. Can we prove, if possible in some quantitative way, that such a function is close (in some suitable sense) to one of the minimizers?

In the latest years several results have been obtained in this direction, showing stability for isoperimetric inequalities [12, 13, 17, 23, 28], the Brunn-Minkowski inequality on convex sets [24], Sobolev [11, 15, 25] and Gagliardo-Nirenberg inequalities [3,15]. We notice that, apart from their own interest, this kind of results have applications in the study of geometric problems (see for instance [9,21,22]) and can be used to obtain quantitative rates of convergence for diffusion equations (see for instance [3]).

Very recently, some quantitative stability results have been proved also for the Brunn-Minkowski inequality on general Borel sets [18,19,24]. The study of this problem involves an interplay between linear structure, analysis, and affine-invariant geometry of Euclidean spaces.

2. Setting and statement of the results

Given two sets $A, B \subset \mathbb{R}^n$, and $c > 0$, we define the set sum and scalar multiple by

$$A + B := \{a + b : a \in A,\ b \in B\}, \quad cA := \{ca : a \in A\} \tag{2.1}$$

Let $|E|$ denote the Lebesgue measure of a set E (if E is not measurable, $|E|$ denotes the outer Lebesgue measure of E). The Brunn-Minkowski inequality states that, given $A, B \subset \mathbb{R}^n$ measurable sets,

$$|A + B|^{1/n} \geq |A|^{1/n} + |B|^{1/n}. \tag{2.2}$$

In addition, if $|A|, |B| > 0$, then equality holds if and only if there exist a convex set $\mathcal{K} \subset \mathbb{R}^n$, $\lambda_1, \lambda_1 > 0$, and $v_1, v_2 \in \mathbb{R}^n$, such that

$$\lambda_1 A + v_1 \subset \mathcal{K}, \quad \lambda_2 B + v_2 \subset \mathcal{K}, \quad |\mathcal{K} \backslash (\lambda_1 A + v_1)| = |\mathcal{K} \backslash (\lambda_2 B + v_2)| = 0.$$

Our aim is to investigate the stability of such a statement.

When $n = 1$, the following sharp stability result holds as a consequence of classical theorems in additive combinatorics (an elementary proof of this result can be given using Kemperman's theorem [7,8]):

Theorem 2.1. *Let $A, B \subset \mathbb{R}$ be measurable sets. If $|A+B| < |A|+|B|+\delta$ for some $\delta \leq \min\{|A|, |B|\}$, then there exist two intervals $I, J \subset \mathbb{R}$ such that $A \subset I$, $B \subset J$, $|I \setminus A| \leq \delta$, and $|J \setminus B| \leq \delta$.*

Concerning the higher dimensional case, in [5,6] M. Christ proved a *qualitative* stability result for (2.2), namely, if $|A + B|^{1/n} - (|A|^{1/n} + |B|^{1/n}) =: \delta \ll 1$ then A and B are close to homothetic convex sets. Since this result relies on compactness, it does not yield any explicit information about the dependence on the parameter δ.

On the *quantitative* side, in [23,24] the author together with F. Maggi and A. Pratelli obtained a sharp stability result for the Brunn-Minkowski inequality on *convex sets*. After dilating A and B appropriately, we can assume $|A| = |B| = 1$ while replacing the sum $A + B$ by a convex combination $S := tA + (1 - t)B$. It follows by (2.2) that $|S| = 1 + \delta$ for some $\delta \geq 0$.

In [23,24] a sharp quantitative stability result is proved when A and B are both convex.

Theorem 2.2. *There is a computable dimensional constant $C_0(n)$ such that if $A, B \subset \mathbb{R}^n$ are convex sets satisfying $|A| = |B| = 1$, $|tA + (1 - t)B| = 1 + \delta$ for some $t \in [\tau, 1 - \tau]$, then, up to a translation,*

$$|A \Delta B| \leq C_0(n) \tau^{-1/2n} \delta^{1/2}$$

(Here and in the sequel, $E \Delta F$ denotes the symmetric difference between two sets E and F, that is $E \Delta F = (E \setminus F) \cup (F \setminus E)$.)

The main theorem in [19] is a quantitative version of Christ's result. Since the proof is by induction on the dimension, it is convenient to allow the measures of $|A|$ and $|B|$ not to be exactly equal, but just close in terms of δ. Here is the main result of that paper.

Theorem 2.3. *Let $n \geq 2$, let $A, B \subset \mathbb{R}^n$ be measurable sets, and define $S := tA + (1 - t)B$ for some $t \in [\tau, 1 - \tau]$, $0 < \tau \leq 1/2$. There are computable dimensional constants N_n and computable functions $M_n(\tau), \varepsilon_n(\tau) > 0$ such that if*

$$\big||A| - 1\big| + \big||B| - 1\big| + \big||S| - 1\big| \leq \delta \tag{2.3}$$

for some $\delta \leq e^{-M_n(\tau)}$, then there exists a convex set $\mathcal{K} \subset \mathbb{R}^n$ such that, up to a translation,

$$A, B \subset \mathcal{K} \qquad \text{and} \qquad |\mathcal{K} \setminus A| + |\mathcal{K} \setminus B| \leq \tau^{-N_n} \delta^{\varepsilon_n(\tau)}.$$

Explicitly, we may take

$$M_n(\tau) = \frac{2^{3^{n+2}} n^{3^n} |\log \tau|^{3^n}}{\tau^{3^n}}, \qquad \varepsilon_n(\tau) = \frac{\tau^{3^n}}{2^{3^{n+1}} n^{3^n} |\log \tau|^{3^n}}.$$

In particular, the measure of the difference between the sets A and B and their convex hull is bounded by a power δ^ϵ, confirming a conjecture of Christ [5].

In order to understand the above statement, it will be useful to go through the conceptual steps that led to its proof.

3. Conceptual path

The question we are trying to address is the following: Assume that (2.2) is almost an equality. Is it true that both A and B are almost convex, and that actually they are close to the same convex set?

Notice that this question has two statements in it:. Indeed, we are wondering if:

- The error in the Brunn-Minkowski inequality controls how far A and B are from their convex hulls **(Convexity)**.
- The error in the Brunn-Minkowski inequality controls the difference between the shapes of A and B **(Homothety)** .

We will proceed by steps as follows: in Section 3.1 we will focus only on the **(Homothety)** issue. More precisely, we assume that A and B are already convex and we prove that, if equality almost holds in (2.2), then A and B have almost the same shape. Then, in Section 3.2 we will focus on the **(Convexity)** issue in the simpler case $A = B$, and we shall prove that A is close to its convex hull. Finally, in Section 3.3 we will deal with the general case.

3.1. Stability on convex sets

Let A, B be bounded convex set with $0 < \lambda \leq |A|, |B| \leq \Lambda$, and set

$$\delta(A, B) := \left|\frac{A+B}{2}\right|^{1/n} - \frac{|A|^{1/n} + |B|^{1/n}}{2}.$$

It follows from (2.2) that $\delta(A, B) \geq 0$, and we would like to show that $\delta(A, B)$ controls some kind of "distance" between the shape of A and the one of B.

In order to compare A and B, we first want them to have the same volume. Hence, we renormalize A so that it has the same measure of B: if $\gamma := \frac{|B|^{1/n}}{|A|^{1/n}}$ then

$$|\gamma A| = |B|.$$

We then define a "distance"[1] between A and B as follows:

$$d(A, B) := \min_{x \in \mathbb{R}^n} |B \Delta (x + \gamma A)|.$$

[1] Notice that d is not properly a distance since it is not symmetric. Still, it is a natural geometric quantity which measures, up to translations, the L^1-closeness between γA and B: indeed, observe that an equivalent formulation for d is

$$d(A, B) := \min_{x \in \mathbb{R}^n} |\|\mathbf{1}_B - \mathbf{1}_{x+\gamma A}\|_{L^1(\mathbb{R}^n)}.$$

The following result has been obtained in [23, Section 4] (see also [24]):

Theorem 3.1. *Let* A, B *be bounded convex set with* $0 < \lambda \leq |A|, |B| \leq \Lambda$. *There exists* $C = C(n, \lambda, \Lambda)$ *such that*

$$d(A, B) \leq C\,\delta(A, B)^{1/2}.$$

As observed in [23, Section 4] the exponent $1/2$ is optimal and the constant C is explicit. The proof of this theorem is obtained by carefully inspecting the proof of Brunn-Minkowski via optimal transport given in [30]. We refer the reader to [20, Section 3] for an idea of the proof.

3.2. Stability when $A = B$

As explained for instance at the beginning of [20, Section 4], the proof of the quantitative stability for Brunn-Minkowski exploiting optimal transportation works only if both A and B are convex. In particular, it cannot be used to solve the **(Convexity)** issue raised at the end of Section 2, and a completely new strategy is needed to address this issue.

The case $n = 1$. Already in the one dimensional case the problem is far from being trivial. Up to rescale A we can always assume that $|A| = 1$. Define

$$\delta_1(A) := |A + A| - 2|A|.$$

It is easy to see that $\delta_1(A)$ cannot control in general $|\mathrm{co}(A) \setminus A|$: indeed take

$$A := [0, 1/2] \cup [L, L + 1/2]$$

with $L \gg 1$. Then

$$A + A = [0, 1] \cup [L, L + 1] \cup [2L, 2L + 1],$$

which implies that $\delta_1(A) = 1 (= |A|)$ while $|\mathrm{co}(A) \setminus A| = L - 1/2$ is arbitrarily large. Luckily, as shown by the following theorem, this is essentially the only thing that can go wrong.

Theorem 3.2. *Let* $A \subset \mathbb{R}$ *be a measurable set with* $|A| = 1$, *and denote by* $\mathrm{co}(A)$ *its convex hull. If* $\delta_1(A) < 1$ *then*

$$|\mathrm{co}(A) \setminus A| \leq \delta_1(A).$$

This theorem can be obtained as a corollary of a result of G. Freiman [26] about the structure of additive subsets of $\mathbb{Z}$. (See [27] or [31, Theorem 5.11] for a statement and a proof.) However, it turns out that to prove of Theorem 3.2 one only needs weaker results, and a simple proof of the above theorem is given in [18, Section 2] (see also [20, Section 4.1]).

The case $n \geq 2$. Let us define the *deficit* of A as

$$\delta(A) := \frac{\left|\frac{1}{2}(A+A)\right|}{|A|} - 1 = \frac{|A+A|}{|2A|} - 1.$$

As mentioned above, one can obtain a precise stability result in one dimension by translating it into a problem on $\mathbb{Z}$. The main result in [18] is a quantitative stability result in arbitrary dimension, showing that a power of $\delta(A)$ dominates the measure of the difference between A and its convex hull $\mathrm{co}(A)$. The proof is done by induction on the dimension, combining several tools from analysis, measure theory, and affine-invariant geometry.

Theorem 3.3. *Let $n \geq 2$. There exist computable dimensional constants $\delta_n, c_n > 0$ such that if $A \subset \mathbb{R}^n$ is a measurable set of positive measure with $\delta(A) \leq \delta_n$, then*

$$\delta(A)^{\alpha_n} \geq c_n \frac{|\mathrm{co}(A) \setminus A|}{|A|}, \qquad \alpha_n := \frac{1}{8 \cdot 16^{n-2} n!(n-1)!}.$$

3.3. Stability when $A \neq B$

As already mentioned in Section 2, when $n = 1$ a sharp stability result holds as a consequence of classical theorems in additive combinatorics.

As in the case $A = B$ and $t = 1/2$ (see Theorem 3.3), the proof of Theorem 2.3 uses the one dimensional result from Theorem 2.1 together with an inductive argument. We want however to point out that, with respect to the one of Theorem 3.3, the proof of Theorem 2.3 is much more elaborate: it combines the techniques of M. Christ in [5,6] with those developed in [18], as well as several new ideas (see [20, Section 5] for a sketch of the proof).

4. Concluding remarks

Although the stability results stated in this note look very much the same (in terms of the statements we want to prove), their proofs involve substantially different methods: Theorem 3.1 relies on optimal transportation techniques, Theorem 3.2 is based on additive combinatorics' arguments, and Theorems 3.3 and 2.3 involve an interplay between measure theory, analysis, and affine-invariant geometry.

We notice the our statements still leave space for improvements: for instance, the exponents α_n and $\beta_n(\tau)$ depend on the dimension, and it looks very plausible to us that they are both non-sharp. An important question in this direction would be to improve our exponents and, if possible, understand what the sharp exponents should be. Notice that this is not a merely academic question, as improving exponents in stability inequalities plays an important role in applications (see for instance [3] and [9]).

It is our belief that this line of research will continue growing in the next years, producing new and powerful stability results.

References

[1] Y. BRENIER, *Polar factorization and monotone rearrangement of vector-valued functions*, Comm. Pure Appl. Math. **44** (1991), no. 4, 375–417.

[2] L. A. CAFFARELLI, *The regularity of mappings with a convex potential*, J. Amer. Math. Soc. **5** (1992), no. 1, 99–104.

[3] E. A. CARLEN and A. FIGALLI, *Stability for a GN inequality and the Log-HLS inequality, with application to the critical mass Keller-Segel equation*, Duke Math. J. **162** (2013), no. 3, 579–625.

[4] E. A. CARLEN and M. LOSS, *Competing symmetries, the logarithmic HLS inequality and Onofri's inequality on $\mathbb{S}^n$*, Geom. Funct. Anal. **2** (1992), 90–104.

[5] M. CHRIST, *Near equality in the two-dimensional Brunn-Minkowski inequality*, Preprint, 2012. Available online at `http://arxiv.org/abs/1206.1965`

[6] M. CHRIST, *Near equality in the Brunn-Minkowski inequality* Preprint, 2012. Available online at `http://arxiv.org/abs/1207.5062`

[7] M. CHRIST, *An approximate inverse Riesz-Sobolev inequality*, Preprint, 2012. Available online at `http://arxiv.org/abs/1112.3715`

[8] M. CHRIST, Personal communication.

[9] M. CICALESE and E. SPADARO, *Droplet minimizers of an isoperimetric problem with long-range interactions*, Comm. Pure Appl. Math. **66** (2013), no. 8, 1298–1333.

[10] D. CORDERO-ERAUSQUIN, B. NAZARET and C. VILLANI, *A mass-transportation approach to sharp Sobolev and Gagliardo-Nirenberg inequalities*, Adv. Math. **182** (2004), no. 2, 307–332.

[11] A. CIANCHI, N. FUSCO, F. MAGGI and A. PRATELLI, *The sharp Sobolev inequality in quantitative form*, J. Eur. Math. Soc. (JEMS) **11** (2009), no. 5, 1105–1139.

[12] A. CIANCHI, N. FUSCO, F. MAGGI and A. PRATELLI, *On the isoperimetric deficit in Gauss space*, Amer. J. Math. **133** (2011), no. 1, 131–186.

[13] M. Cicalese and G. P. Leonardi, *A selection principle for the sharp quantitative isoperimetric inequality*, Arch. Ration. Mech. Anal. **206** (2012), no. 2, 617–643.

[14] M. Del Pino and J. Dolbeault, *Best constants for Gagliardo-Nirenberg inequalities and applications to nonlinear diffusions*, J. Math. Pures Appl. **81** (2002), no. 9, 847–875.

[15] J. Dolbeault and G. Toscani, *Improved interpolation inequalities, relative entropy and fast diffusion equations*, Ann. Inst. H. Poincaré Anal. Non Linéaire **30** (2013), no. 5, 917–934.

[16] L. Esposito, N. Fusco and C. Trombetti, *A quantitative version of the isoperimetric inequality: the anisotropic case*, Ann. Sc. Norm. Super. Pisa Cl. Sci. (5) **4** (2005) no. 4, 619–651.

[17] A. Figalli and E. Indrei, *A sharp stability result for the relative isoperimetric inequality inside convex cones*, J. Geom. Anal. **23** (2013), no. 2, 938–969.

[18] A. Figalli and D. Jerison, *Quantitative stability for sumsets in* $\mathbb{R}^n$, J. Eur. Math. Soc. (JEMS), 2012.

[19] A. Figalli and D. Jerison, *Quantitative stability for the Brunn-Minkowski inequality*, preprint, 2013.

[20] A. Figalli, *Stability results for the Brunn-Minkowski inequality*, Proceeding ICM 2014, to appear.

[21] A. Figalli and F. Maggi, *On the shape of liquid drops and crystals in the small mass regime*, Arch. Ration. Mech. Anal. **201** (2011), no. 1, 143–207.

[22] A. Figalli and F. Maggi, *On the isoperimetric problem for radial log-convex densities*, Calc. Var. Partial Differential Equations **48** (2013), no. 3-4, 447–489.

[23] A. Figalli, F. Maggi and A. Pratelli, *A mass transportation approach to quantitative isoperimetric inequalities*, Invent. Math. **182** (2010), no. 1, 167–211.

[24] A. Figalli, F. Maggi and A. Pratelli, *A refined Brunn-Minkowski inequality for convex sets*, Ann. Inst. H. Poincaré Anal. Non Linéaire **26** (2009), no. 6, 2511–2519.

[25] A. Figalli, F. Maggi and A. Pratelli, *Sharp stability theorems for the anisotropic Sobolev and log-Sobolev inequalities on functions of bounded variation*, Adv. Math. **242** (2013), 80–101.

[26] G. A. Freiman, *The addition of finite sets*, I. (Russian) Izv. Vyss. Ucebn. Zaved. Matematika **13** (1959), no. 6, 202–213.

[27] G. A. Freiman, "Foundations of a Structural Theory of Set Addition", Translated from the Russian. Translations of Mathematical Monographs, Vol. 37, American Mathematical Society, Providence, R. I., 1973.

[28] N. Fusco, F. Maggi and A. Pratelli, *The sharp quantitative isoperimetric inequality*, Ann. of Math. (2) **168** (2008), 941–980.
[29] R. J. GARDNER, *The Brunn-Minkowski inequality*, Bull. Amer. Math. Soc. (N.S.) **39** (2002), no. 3, 355–405.
[30] R. J. MCCANN, *A convexity principle for interacting gases*, Adv. Math. **128** (1997), no. 1, 153–179.
[31] T. TAO and V. VU, "Additive Combinatorics", Cambridge Studies in Advanced Mathematics, 105, Cambridge University Press, Cambridge, 2006.

Schanuel's Conjecture: algebraic independence of transcendental numbers

Michel Waldschmidt

Abstract. Schanuel's conjecture asserts that given linearly independent complex numbers $x_1, ..., x_n$, there are at least n algebraically independent numbers among the $2n$ numbers

$$x_1, \dots, x_n, \; e^{x_1}, \dots, e^{x_n}.$$

This simple statement has many remarkable consequences; we explain some of them. We also present the state of the art on this topic.

1. The origin of Schanuel's Conjecture

To prove that a constant arising from analysis is irrational is most often a difficult task. It was only in 1873 that C. Hermite succeeded to prove the transcendence of $e = 2.718\,281\dots$ and it took 9 more years before F. Lindemann obtained the transcendence of $\pi = 3.141\,592\dots$, thereby giving a final negative solution to the Greek problem of squaring the circle. This method produces the so–called Hermite–Lindemann Theorem, which states that *for any nonzero complex number z, one at least of the two numbers z, e^z is transcendental.*

To prove algebraic independence of transcendental numbers is much harder and few results are known. The earliest one is the Lindemann–Weierstrass Theorem, which states that *for **Q**–linearly independent algebraic numbers $\beta_1, \dots, \beta_n$, the numbers $e^{\beta_1}, \dots, e^{\beta_n}$ are algebraically independent.* This is one of the very few statements on algebraic independence of numbers related with the exponential function. Even the quest of a conjectural general statement has been a challenge for many years. A.O. Gel'fond made an attempt in a one page note in the Comptes-Rendus de l'Académie des Sciences de Paris in 1934 (see the appendix), just after he solved the 7th problem of Hilbert on the transcendence of α^β, a problem which was solved by Th Schneider, at the same time, with a different but similar method.

Fourteen years later, A.O. Gel'fond was able to prove a very special case of the first theorem of his note, when he proved that *the two numbers $2^{\sqrt[3]{2}}$ and $2^{\sqrt[3]{4}}$ are algebraically independent.* This proof is a master piece, which paved the way for a number of later developments (see Section 2). To find "the right" conjectural statement took 15 more years, until S. Schanuel had a remarkable insight:

Schanuel's Conjecture. *Let $x_1, ..., x_n$ be* **Q**–*linearly independent complex numbers. Then, among the* $2n$ *numbers*

$$x_1, \ldots, x_n, \; e^{x_1}, \ldots, e^{x_n},$$

there are at least n algebraically independent numbers.

This statement was proposed by S. Schanuel during a course given by S. Lang at Columbia in the 60's. The reference is Lang's book *Introduction to transcendental numbers*, Addison-Wesley 1966.

2. Related results

A version of Schanuel's conjecture for power series over $\mathbb{C}$ has been proved by J. Ax in 1968 using differential algebra. In his talk at ICM 1970 in Nice *Transcendence and differential algebraic geometry*, Ax describes his contribution in a general setting, where he also quotes Brun's Theorem on the integrals of the three body problem. The stronger version for power series over the field of algebraic numbers is due to R. Coleman (1980), who combined the ideas of Ax with p–adic analysis and the Čebotarev density Theorem.

A former student of Schanuel, W.D. Brownawell, obtained an elliptic analog of Ax's Theorem in a joint work with K. Kubota.

More recently, deep connections between Schanuel's Conjecture and model theory have been investigated by E. Hrushovski, B. Zilber, J. Kirby, A. Macintyre, D.E. Marker, G. Terzo, A.J. Wilkie, D. Bertrand and others.

Under the assumption of Schanuel's Conjecture, "most often", the transcendence degree is $2n$. Indeed, the set of tuples $(x_1, \ldots, x_n)$ in $\mathbb{C}^n$ such that the $2n$ numbers

$$x_1, \ldots, x_n, \; e^{x_1}, \ldots, e^{x_n}$$

are algebraically independent is a G_δ set (countable intersection of dense open sets) in Baire's classification (a *generic set* for dynamical systems) and has full Lebesgue measure. However, this result is not very significant, since it is true for any transcendental function in place of the exponential function.

The transcendence degree can also be as small as n, for instance, when the x_i are algebraic (Lindemann–Weierstrass Theorem), or when the e^{x_i} are algebraic (algebraic independence of logarithms of algebraic numbers

— see Section 3) and also when, for each i, either x_i or e^{x_i} is algebraic, like in the first statement of Gel'fond's 1934 note.

With K. Senthil Kumar and R. Thangadurai, we recently proved that *given two integers m and n with $1 \leq m \leq n$, there exist uncountably many tuples $(x_1, \dots, x_n)$ in $\mathbb{R}^n$ such that $x_1, \dots, x_n$ and $e^{x_1}, \dots, e^{x_n}$ are all Liouville numbers and the transcendence degree of the field*

$$\mathbf{Q}(x_1, \dots, x_n, \ e^{x_1}, \dots, e^{x_n})$$

is $n + m$. Whether such a result holds in case $m = 0$ is unclear: for instance (with $n = 1$) we do not even know whether there are Liouville numbers x such that e^x is also a Liouville number and the two numbers x and e^x are algebraically dependent.

For $n = 1$, Schanuel's Conjecture is nothing else than the Hermite–Lindemann Theorem (see Section 1).

For $n = 2$, Schanuel's Conjecture is not yet known:

Schanuel's Conjecture for $n = 2$: *If x_1, x_2 are* **Q***–linearly independent complex numbers, then among the 4 numbers $x_1, x_2, e^{x_1}, e^{x_2}$, at least two are algebraically independent.*

A few consequences are the following open problems:

- With $x_1 = 1, x_2 = i\pi$: *the number e and π are algebraically independent.*
- With $x_1 = 1, x_2 = e$: *the number e and e^e are algebraically independent.*
- With $x_1 = \log 2, x_2 = (\log 2)^2$: *the number* $\log 2$ *and* $2^{\log 2}$ *are algebraically independent.*
- With $x_1 = \log 2$, $x_2 = \log 3$: *the number* $\log 2$ *and* $\log 3$ *are algebraically independent.*

Among many mathematicians who contributed to prove partial results in the direction of Schanuel's Conjecture are Ch. Hermite, F. Lindemann, C. L. Siegel, A. O. Gel'fond, Th. Schneider, A. Baker, S. Lang, W. D. Brownawell, D. W. Masser, D. Bertrand, G. V. Chudnovsky, P. Philippon, G. Diaz, G. Wüstholz, Yu. V. Nesterenko, D. Roy…

An important step, already mentioned in Section 1, is due to A. O. Gel'fond who proved in 1948 that *the two numbers $2^{\sqrt[3]{2}}$ and $2^{\sqrt[3]{4}}$ are algebraically independent.* More generally, he proved that *if α is an algebraic number, $\alpha \neq 0$, $\log \alpha \neq 0$ and if β is an algebraic number of degree $d \geq 3$, then two at least of the numbers*

$$\alpha^{\beta}, \ \alpha^{\beta^2}, \ \dots, \alpha^{\beta^{d-1}}$$

are algebraically independent. Recall that $\alpha^{\beta} = \exp(\beta \log \alpha)$.

The deep method devised by Gel'fond was extended by G.V. Chudnovsky (1978). One of the most remarkable results by Chudnovsky states that π *and* $\Gamma(1/4) = 3.625\,609\ldots$ *are algebraically independent*. Also π *and* $\Gamma(1/3) = 2.678\,938\ldots$ *are algebraically independent*. Until then, the transcendence of $\Gamma(1/4)$ and $\Gamma(1/3)$ was not known. The next important step is due to Yu.V.Nesterenko (1996) who proved the algebraic independence of $\Gamma(1/4), \pi = 3.141\,592$ and $e^\pi = 23.140\,692\ldots$ Until then, the algebraic independence of π and e^π was not yet known. Nesterenko's proof uses modular functions.

The number e^π conceals mysteries.

Open problem: prove that the number e^π is not a Liouville number: *there exists a positive absolute constant* κ *such that for any* $p/q \in \mathbf{Q}$ *with* $q \geq 2$,

$$\left| e^\pi - \frac{p}{q} \right| > \frac{1}{q^\kappa}.$$

An other open problem, consequence of Schanuel's Conjecture, is the algebraic independence of the three numbers e, π and e^π. More generally, according to Schanuel's Conjecture, the following numbers are algebraically independent (none of them is known to be irrational!):

$$e + \pi,\ e\pi,\ \pi^e,\ e^{\pi^2},\ e^e, e^{e^2}, \ldots,\ e^{e^e}, \ldots,\ \pi^\pi, \pi^{\pi^2}, \ldots\ \pi^{\pi^\pi}\ \ldots$$

$$\log \pi,\ \log(\log 2),\ \pi \log 2,\ (\log 2)(\log 3),\ 2^{\log 2},\ (\log 2)^{\log 3}\ \ldots$$

The proof is an easy exercise using Schanuel's Conjecture. The list of similar exercises is endless, some recent papers pursue this direction with no new idea. A less trivial result has been proved in a joint paper in 2008 by J. Bober, C. Cheng, B. Dietel, M. Herblot, Jingjing Huang, H. Krieger, D. Marques, J. Mason, M. Mereb and R. Wilson. Define $E_0 = \overline{\mathbf{Q}}$. Inductively, for $n \geq 1$, define E_n as the algebraic closure of the field generated over E_{n-1} by the numbers $\exp(x) = e^x$, where x ranges over E_{n-1}. Let E be the union of $E_n, n \geq 0$. In a similar way, define $L_0 = \overline{\mathbf{Q}}$. Inductively, for $n \geq 1$, define L_n as the algebraic closure of the field generated over L_{n-1} by the numbers y, where y ranges over the set of complex numbers such that $e^y \in L_{n-1}$. Let L be the union of $L_n, n \geq 0$. Then Schanuel's Conjecture implies that *the fields E and L are linearly disjoint over* $\overline{\mathbf{Q}}$. As a consequence, π does not belong to E, a statement proposed by S. Lang.

3. Algebraic independence of logarithms of algebraic numbers

Denote by $\mathcal{L}$ the set of complex numbers λ for which e^λ is algebraic. This set $\mathcal{L}$ is the $\mathbf{Q}$–subspace of $\mathbb{C}$ of all logarithms of nonzero algebraic numbers:

$$\mathcal{L} = \{\log \alpha \mid \alpha \in \overline{\mathbf{Q}}^{\times}\}.$$

Arguably, the most important special case of Schanuel's Conjecture is:

Conjecture (Algebraic independence of logarithms of algebraic numbers). *Let $\lambda_1, \ldots, \lambda_n$ be* **Q***-linearly independent elements in $\mathcal{L}$. Then the numbers $\lambda_1, \ldots, \lambda_n$ are algebraically independent over* **Q**.

The homogeneous version is often sufficient for applications:

Conjecture (Homogeneous algebraic independence of logarithms of algebraic numbers). *Let $\lambda_1, \ldots, \lambda_n$ be* **Q***-linearly independent elements in $\mathcal{L}$. Let $P \in \mathbf{Q}[X_1, \ldots, X_n]$ be a* homogeneous *nonzero polynomial. Then*

$$P(\lambda_1, \ldots, \lambda_n) \neq 0.$$

In 1968, A. Baker proved that *if $\lambda_1, \ldots, \lambda_n$ are* **Q***–linearly independent logarithms of algebraic numbers, then the numbers $1, \lambda_1, \ldots, \lambda_n$ are linearly independent over the field $\overline{\mathbf{Q}}$ of algebraic numbers.*

Baker's Theorem is a special case of Schanuel's Conjecture: while Schanuel's Conjecture deals with algebraic independence (over **Q** or over $\overline{\mathbf{Q}}$, it is the same), Baker's Theorem deals with linear independence over $\overline{\mathbf{Q}}$.

An open problem is to prove that the transcendence degree over **Q** of the field $\mathbf{Q}(\mathcal{L})$ generated by all the logarithms of nonzero algebraic numbers is at least 2. However, even if the answer is not yet known, this does not mean that nothing is known: partial results have been proved, in particular by D. Roy, thanks to a reformulation of the problem. Instead of taking logarithms of algebraic numbers and looking for the algebraic independence relations, he fixes a polynomial and looks at the points, with coordinates logarithms of algebraic numbers, on the corresponding hypersurface. One easily checks that the homogeneous conjecture on algebraic independence of logarithms of algebraic numbers is equivalent to:

Conjecture (D. Roy). *For any algebraic subvariety X of $\mathbb{C}^n$ defined over the field $\overline{\mathbf{Q}}$ of algebraic numbers, the set $X \cap \mathcal{L}^n$ is the union of the sets $\mathcal{V} \cap \mathcal{L}^n$, where $\mathcal{V}$ ranges over the set of vector subspaces of $\mathbb{C}^n$ which are contained in X.*

Special cases of this statement have been proved by D. Roy and S. Fischler.

Even the nonexistence of quadratic relations among logarithms of algebraic numbers is not proved. For instance, Schanuel's Conjecture implies that a relation like

$$\log \alpha_1 \log \alpha_2 = \log \alpha_3$$

among nonzero logarithms of algebraic numbers is not possible. A special case would be the transcendence of the number e^{π^2} – it is not yet proved that this number is irrational.

A special case of the nonexistence of nontrivial homogeneous quadratic relations between logarithms of algebraic numbers is the four exponentials conjecture, which occurs in some works on highly composite numbers by S. Ramanujan, L. Alaoglu and P. Erdős. A very special case, which is open yet, is to prove that *if t is a real number such that 2^t and 3^t are integers, then t is an integer.*

The homogeneous conjecture of algebraic independence of logarithms of algebraic numbers can be stated in an equivalent way as saying that, under suitable assumptions, the determinant of a square matrix having entries in $\mathcal{L}$ does not vanish. To deduce such a statement from Schanuel's Conjecture is easy, the converse relies on the fact, proved by D. Roy, that *any polynomial in variables $X_1, \ldots, X_n$ is the determinant of a matrix having entries which are linear forms in $1, X_1, \ldots, X_n$.*

4. Further consequences of Schanuel's Conjecture

In 1979, P. Bundschuh investigated the transcendence of numbers of the form

$$\sum_{n \geq 1} \frac{A(n)}{B(n)},$$

where $A/B \in \mathbf{Q}(X)$ with $\deg B \geq \deg A + 2$. As an example, he noticed that

$$\sum_{n=0}^{\infty} \frac{1}{n^2+1} = \frac{1}{2} + \frac{\pi}{2} \cdot \frac{e^{\pi} + e^{-\pi}}{e^{\pi} - e^{-\pi}} = 2.076\,674\,047\,4\ldots$$

while

$$\sum_{n=0}^{\infty} \frac{1}{n^2-1} = \frac{3}{4}$$

(telescoping series).

From the Theorem of Nesterenko it follows that the number

$$\sum_{n=2}^{\infty} \frac{1}{n^s - 1}$$

is transcendental over $\mathbf{Q}$ for $s = 4$. The transcendence of this number for even integers $s \geq 4$ would follow as a consequence of Schanuel's

Conjecture. The example $A(X)/B(X) = 1/X^3$ shows that it will be hard to achieve a very general result, since :

$$\zeta(3) = \sum_{n\geq 1} \frac{1}{n^3},$$

an irrational number not yet known to be transcendental. Such series of values of rational fractions were studied later by S.D. Adhikari, R. Tijdeman, T.N. Shorey, R. Murty, C. Weatherby and others.

An important recent development is due to the work of R. Murty and several of his collaborators, including K. Murty, N. Saradha, S. Gun, P. Rath, C. Weatherby... They use Schanuel's conjecture to study not only the arithmetic nature of numbers like Euler's constant, Catalan's Constant, values of Euler Gamma function, the digamma function and Barnes's multiple Gamma function, but also the non vanishing of L–series at critical points.

5. Roy's program towards Schanuel's Conjecture

In the Journées Arithmétiques in Roma 1999, D. Roy revealed his ambitious program to prove Schanuel's Conjecture. So far, this is the only approach which is known toward a proof of Schanuel's Conjecture. D. Roy introduces a new conjecture of his own, which bears some similarity with known criteria of algebraic independence and he proves that his new conjecture is equivalent to Schanuel's Conjecture. Both sides of the proof of equivalence are difficult and involve a clever use of the transcendence machinerie.

Let $\mathcal{D}$ denote the derivation

$$\mathcal{D} = \frac{\partial}{\partial X_0} + X_1 \frac{\partial}{\partial X_1}$$

over the ring $\mathbb{C}[X_0, X_1]$. The *height* of a polynomial $P \in \mathbb{C}[X_0, X_1]$ is defined as the maximum of the absolute values of its coefficients.

Roy's Conjecture. *Let k be a positive integer, $y_1, \ldots, y_k$ complex numbers which are linearly independent over $\mathbf{Q}$, $\alpha_1, \ldots, \alpha_k$ non-zero complex numbers and s_0, s_1, t_0, t_1, u positive real numbers satisfying*

$$\max\{1, t_0, 2t_1\} < \min\{s_0, 2s_1\}$$

and

$$\max\{s_0, s_1 + t_1\} < u < \frac{1}{2}(1 + t_0 + t_1).$$

Assume that, for any sufficiently large positive integer N, there exists a non-zero polynomial $P_N \in \mathbb{Z}[X_0, X_1]$ with partial degree $\le N^{t_0}$ in X_0, partial degree $\le N^{t_1}$ in X_1 and height $\le e^N$ which satisfies

$$\left|(\mathcal{D}^k P_N)\Big(\sum_{j=1}^{k} m_j y_j, \prod_{j=1}^{k} \alpha_j^{m_j}\Big)\right| \le \exp(-N^u)$$

for any non-negative integers k, $m_1, \ldots, m_k$ with $k \le N^{s_0}$ and $\max\{m_1, \ldots, m_k\} \le N^{s_1}$. Then

$$\operatorname{tr\,deg}\mathbf{Q}(y_1, \ldots, y_k, \alpha_1, \ldots, \alpha_k) \ge k.$$

D. Roy already obtained partial results for the groups $\mathbf{G_a}$ and $\mathbf{G_m}$; recently he reached the first result for $\mathbf{G_a} \times \mathbf{G_m}$, so far only when the subset is reduced to a single point.

Roy's Conjecture depends on parameters s_0, s_1, t_0, t_1, u in a certain range. He proved that if his conjecture is true for one choice of values of these parameters in the given range, then Schanuel's Conjecture is true and that conversely, if Schanuel's Conjecture is true, then his conjecture is true for all choices of parameters in the same range. Recently, Nguyen Ngoc Ai Van extended the range of these parameters.

6. Ubiquity of Schanuel's Conjecture

Schanuel's Conjecture occurs in many different places; most often only the special case of homogeneous algebraic independence of logarithms of algebraic numbers is required. Some regulators are determinant of matrices with entries logarithms of algebraic numbers; the fact that they do not vanish means that some algebraic relation between these logarithms is not possible. An example, involving the p–adic analog of Schanuel's Conjecture, is Leopoldt's Conjecture on the p–adic rank of the units of an algebraic number field. The nondegenerescence of heights is also sometimes a consequence of Schanuel's Conjecture, as shown by D. Bertrand. Some special cases of a conjecture of B. Mazur on the density of rational points on algebraic varieties can be deduced from Schanuel's Conjecture. Other applications are related with questions on algebraic tori in the works of D. Prasad and of G. Prasad.

Another far-reaching topic is the connection between Schanuel's Conjecture and other conjectures on transcendental number theory, including the Conjecture of A. Grothendieck on the periods of abelian varieties, the conjecture due to Y. André on motives and the Conjecture of M. Kontsevich and D. Zagier on periods.

Appendix

Comptes rendus hebdomadaires des séances de l'Académie des sciences, Gauthier-Villars (Paris) **199** (1934), p. 259. http://gallica.bnf.fr/

SÉANCE DU 23 JUILLET 1934

ARITHMÉTIQUE. — Sur quelques résultats nouveaux dans la théorie des nombres transcendants.
Note de M. **A. Gelfond**, présentée par M. Hadamard.

J'ai démontré [1] que le nombre ω^r, où $\omega \neq 0, 1$ est un nombre algébrique et r un nombre algébrique irrationnel, doit être transcendant.

Par une généralisation de la méthode qui sert pour la démonstration du théorème énoncé, j'ai démontré les résultats plus généraux suivants :

I. THÉORÈME. — *Soient $P(x_1, x_2, \dots, x_n, y_1, \dots, y_m)$ un polynôme à coefficients entiers rationnels et $\alpha_1, \alpha_2, \dots, \alpha_n, \beta_1, \dots, \beta_m$ des nombres algébriques, $\beta_i \neq 0, 1$.*
L'égalité

$$P(e^{\alpha_1}, e^{\alpha_2}, \dots, e^{\alpha_n}, \ln \beta_1, \dots, \ln \beta_m) = 0$$

est impossible ; les nombres, $\alpha_1, \alpha_2, \dots, \alpha_n$, et aussi les nombres $\ln \beta_1, \dots, \ln \beta_m$ sont linéairement indépendants dans le corps des nombres rationnels.

Ce théorème contient, comme cas particuliers, le théorème de Hermite et Lindemann, la résolution complète du problème de Hilbert, la transcendance des nombres $e^{\omega_1 e^{\omega_2}}$ (où ω_1 et ω_2 sont des nombres algébriques), le théorème sur la transcendance relative des nombres e et π.

II. THÉORÈME. — *Les nombres*

$$e^{\omega_1 e^{\omega_2 e^{\cdots \omega_{n-1} e^{\omega_n}}}} \quad \textit{et} \quad \alpha_1^{\alpha_2^{\alpha_3^{\cdots \alpha_m}}},$$

où $\omega_1 \neq 0, \omega_2, \dots, \omega_n$ et $\alpha_1 \neq 0, 1$, $\alpha_2 \neq 0, 1$, $\alpha_3 \neq 0$, $\alpha_4, \dots, \alpha_m$ sont des nombres algébriques, sont des nombres transcendants et entre les nombres de cette forme n'existent pas de relations algébriques, à coefficients entiers rationnels (non triviales).

La démonstration de ces résultats et de quelques autres résultats sur les nombres transcendants sera donnée dans un autre Recueil.

---

[1] *Sur le septième problème de D. Hilbert (C.R. de l'Acad. des Sciences de l'U.R.S.S.*, 2, I, 1er avril 1934, et *Bull. de l'Acad. des Sciences de l'U.R.S.S.*, 7e série, 4, 1934, p. 623).

COLLOQUIA

The volumes of this series reflect lectures held at the "Colloquio De Giorgi" which regularly takes place at the Scuola Normale Superiore in Pisa. The Colloquia address a general mathematical audience, particularly attracting advanced undergraduate and graduate students.

Published volumes

1. Colloquium De Giorgi 2006. ISBN 978-88-7642-212-6
2. Colloquium De Giorgi 2007 and 2008. ISBN 978-88-7642-344-4
3. Colloquium De Giorgi 2009. ISBN 978-88-7642-388-8, e-ISBN 978-88-7642-387-1
4. Colloquium De Giorgi 2010–2012. ISBN 978-88-7642-455-7, e-ISBN 978-88-7642-457-1
5. Colloquium De Giorgi 2013 and 2014. ISBN 978-88-7642-514-1, e-ISBN 978-88-7642-515-8

Fotocomposizione CompoMat srl, Loc. Braccone, 02040 Configni (RI)
Finito di stampare nel mese di febbraio 2015
dalla CSR srl, Via di Pietralata 157, 00158 Roma